If You Have Flown

There are no words that can express
The magic of that wilderness
That wilderness away up high
Where banks of clouds float softly by
And hide the problems of earth below.
But then, you know —
If you have flown.

If you have flown, then you know
The beauty of the world below —
The meadow green, the waters sapphire blue,
You've felt that it belonged alone to you.
And as your ship obeyed your slightest will
You've felt a thrill —
If you have flown.

To those who sail the sky above
Comes peace of mind and understanding love.
As gently earth and clouds drift by.
All is beautiful, serene.
You know exactly what I mean —
If you have flown.

~ Betty Huyler Gillies
1908 - 1998

THIS FLYING LIFE

By

Claire L. Walters

and

Betty McMillen Loufek

This book is dedicated to

*All the flight instructors
in all the world,
past, present and future;
"Let such teach, who
themselves excel"
(Pope, 1711)*

*For you are the
heart and soul of
this flying life.*

Copyright © 1999
Betty McMillen Loufek and Claire L. Walters

ISBN 0-9670656-0-7

Library of Congress Catalogue Card Number: 99-72379

To order additional copies of this book, contact:
AirWoman Press
P.O. Box 721
Camarillo, CA 93011-0721

728 Hunt Circle
Camarillo, CA 93012

**— *An AirWoman Press Publication* —
*Camarillo, California***

Printed in the United States of America by
Maverick Publications, Inc. • P.O. Box 5007 • Bend, OR 97708

Acknowledgments

Mrs. Gordon J. Frey (Marjorie) read the book in manuscript. Her enthusiastic remarks helped keep up my creative fires. She was married nearly 46 years to a Navy pilot who in WWII flew the route to Hawaii and on to the South Pacific many times, and later was a flight instructor for the Navy. Marjorie flew with him on many private flights, both in open cockpits and in closed cabins. As she read the manuscript, she wished Gordon was still there with her reading the book, too; she is sure he would have enjoyed it. Thank you, Marj, for all your kind words and help.

Mr. T. Russell Wingate was a most helpful critic. He caught what would have been embarrassing errors, suggested ways to make the book more readable to non-aviation-minded readers.

Mrs. Betty Gillies, whose poem "If You have Flown", is at the front of this book, was the brilliant chairman of the All Woman Transcontinental Air Race (Powder Puff Derby) in its formative years. Her firm hand and stated philosophy was the compass that kept it on its 30-year course as the finest air race ever flown.

Monie Pease, whose high energy helps drive us all on. She successfully nominated Claire for the prestigious NAA Elder Statesman Award in 1998. May her flight records stand forever!

Lota E. Blythe, Santa Ana High School creative writing teacher, encouraged us to go beyond our dreams.

Robert Farrar, Santa Ana High School principal, encouraged us. His brother, Les Farrar, hired Claire for her first commercial flying, ferrying aircraft.

Beulah Davis McMillen, our mother, encouraged our flying and writing ambitions.

My son David moved me into the computer world and provides assistance whenever I need it. Thanks, Dave, I could not have finished this without you.

My son Richard and wife Gretchen, and my two wonderful grandchildren, Whitney and Killarney. All sailing champions. Thanks for cheering me on.

Holly Beth Mingo and Christopher Mingo, my grandchildren in Denmark, want to read the book. Can't ask for better than that.

Peggy Ketteman, my editor and book designer at Maverick Publications, has done a tremendous job. Thank you.

Thanks to all our friends who you will meet in this book. They tolerated — or kindly pretended to — my intrusion into their lives asking for anecdotes and photographs.

Betty Loufek

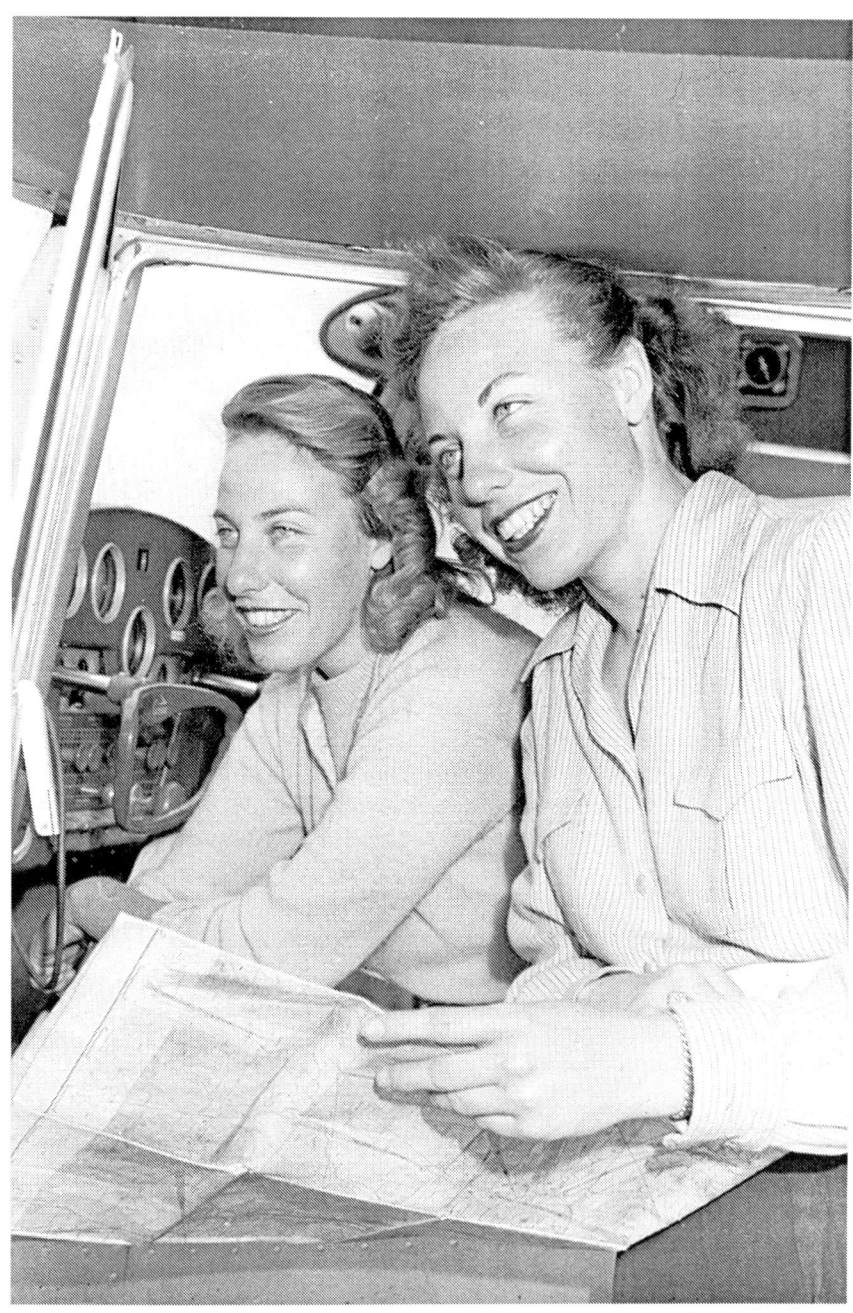

Claire Walters is on the right foreground; her twin sister, Betty Loufek, is on the left. They were preparing for the second Powder Puff Derby in 1948.

Photo courtesy of Dee Thurmond; ©1999 Dee Thurmond

Chapter 1

My twin sister Betty and I were born February 20, 1924, in Santa Ana, California, to Paul and Beulah McMillen. I was the elder by ten minutes. I lived there until the fall of 1943 when I left for Quartzsite Airport in the Arizona desert.

Our parents were hard-working Kansans, born on farms before the turn of the twentieth century. After their marriage, they moved to Oklahoma and Nevada farms, then to Bend, Oregon, to work for the railroad, and finally, to Southern California.

Mother, born in 1890, was the eldest of 12 children. A school was near enough to walk to; she completed the eighth grade before she left home at sixteen. Dad, born in 1888, finished the sixth grade, enough in those days for a farmboy, but he knew the value of education and made sure all his children graduated from high school before letting any of us off the leash. This was the main reason, I think, for coming to California. Other than the weather, of course.

Mother was extremely capable, as any woman of pioneer stock had to be to keep her family safe from the many dangers: human and animal predators, farm machinery, sudden and deadly changes in the weather, and terrible illnesses for which there was no medicine.

Mom could shoot a rifle, ride horses bareback, hitch up and drive a wagon behind four horses, cook meals for her family and the many field hands at planting and harvest times, chop the wood, feed the chickens, keep the vegetable garden.

She was flexible and eager to try new things. She learned to drive as soon as they had a car; when I became a licensed pilot, she flew with me any time I could take her.

Dad learned carpentry and cement work, and when they moved to Santa Ana, he became a cement contractor and did very well until the Great Depression came in 1929. The banks refused to renew the one-year and five-year mortgages on homes, and they lost theirs along with so many others. Dad never recovered from that.

Santa Ana's aviation history began early. It was there in 1909 that Glenn L. Martin started his aircraft company. Two years later he moved it to Los Angeles, received aviator certificate number 56, and in 1913 he delivered his first Army trainer. William E. Boeing, Lawrence D. Bell and Donald W. Douglas were among those employed later by Martin.

In the early years of the 20th century, aviation feats were big news, splashed on the front pages of newspapers and blared over the radios.

In March 1924, a month after I was born, four Douglas airplanes took off to fly around the world. Two completed the circuit in about sixteen days. In June 1924, a dawn-to-dusk flight in one day was made from New York to San Francisco.

Many distance, altitude, and endurance flights were quickly set and quickly broken again and again in the United States and throughout the world by many fliers.

Some were military exploits, but most were by private citizens intent on developing aircraft capabilities and expanding airmail and passenger routes.

We were three when Charles Lindbergh flew nonstop from New York to Paris; we were four when Amelia Earhart flew as a passenger across the Atlantic. She flew it alone in 1932 when we were eight; we were thirteen when Amelia was lost in the Pacific.

In those years and since, many women made flights of all kinds. I know a few of these women, now in their 80s and 90s, and they still talk enthusiastically about flying; they had such fun. These pioneers helped make aviation grow and mature, and in a less flamboyant way so do my flying friends and I today.

My generation was enthralled with aviation. Thousands learned to fly, many became aeronautical engineers, others created the metals and fabrics or invented instruments and navigational aids. In just 60 years, aviation grew from the first solo power flights to carrying millions of people all around our earth, to putting the men on the moon.

When we were three years old, Dad gave us each a box of real tools and we promptly made a model airplane from pine wood. Not a good one, but it had a wing and a fuselage! An airplane, but too heavy and it wouldn't fly. I put all my dolls in my wagon and went through the neighborhood trying to sell them so I could buy an airplane that would fly.

By the sixth grade we had read all the books on aviation in our school library. My goals expanded and I knew I wanted to be a flight instructor and own a flight school.

When we were 13 we lived for awhile on a two-acre farm near Garden Grove, California. We earned enough from raising corn and a pig to pay for an airplane ride in a cabin Stinson flown by one of the Martin brothers. I loved it; my future was set.

When we were 15 Mom and Dad drove us at night to the weekly ground school at Eddie Martin's Airport, a grass field near where

Orange County Airport (John Wayne) is now. We studied traffic rules of the air, navigation, meteorology, and flight theory. Everyone in the class was nice to us and we were in our heaven, on an airport talking to fliers.

In my senior year in high school, I was asked by the school counselor why I was taking physics and math. There was only one other girl in the large class. She told me I should be taking the secretarial course (called "commercial course" then). I told the counselor, "I want to fly and I need those courses."

"You're poor. It costs lots of money to fly. You don't have any."

"I will be a flight instructor and own a flight school."

She tried to talk me out of it. I told her that I believed I could be anything I wanted to be if I wanted it enough to work for it. In English class we had read Henry Thoreau's "Walden" and I took one message to heart: "... if one advances confidently in the direction of his dreams, and endeavors to live the life which he has imagined, he will meet with a success unexpected in common hours... If you have built castles in the air, your work need not be lost; that is where they should be. Now put the foundations under them."

I knew that to become a flight instructor required a known series of steps which I could do as I earned the money for them. Each step required serious dedication to complete.

I told her there were women flight instructors. I was going to be one of them. And, I was going to own my own school, ferry planes, be in air races, and fly all over the world.

Just 11 years later, at the testimonial dinner held for me at my college after I had won the great transcontinental air race, the Powder Puff Derby, with my co-pilot Frances Bera, my counselor apologized and said she was proud of my success.

Unfortunately, her attitude was one held by counselors at most schools at the time, and there are too many yet today. They stereotyped women and discouraged all but the most stubborn from taking courses involving science and math. Courses needed for engineering, medicine, and so many other careers. This misguided counseling blocked women from entering many university programs. In later years, when I spoke at universities, I stressed this.

We need more counselors like Amelia Earhart at Purdue University just before her last flight in 1937. She counseled women students and many of them reached further than they ever thought possible. Some years ago, I met a dynamo woman, a retired aeronautical engineer now in her 80s. She told me she had talked to Earhart for just a few moments, told her she wanted to learn to fly.

Earhart said that was a good idea, but why not also become an aeronautical engineer?

We graduated from high school in June, 1941. My sister and I hunted for jobs. A cannery near Fullerton Airport was hiring unskilled labor. We caught the intercity bus and got off near the cannery. We filled out the applications and waited in the "bull pen," an area fifty feet from the factory building and separated from it by a high wire fence with a gate. A guard stood at the gate. Those who wanted jobs waited in the bull pen.

About 50 persons sat on unpainted wooden benches. Most of them had fled the Midwest dust bowl. They had little formal schooling but they knew the practical things that we didn't: what each job entailed, those that were dangerous. The men received 45 cents an hour; women, 25 to 35 cents. Five cents more for dangerous jobs.

We sat for several days on those benches, learning from those around us how to get a job. When a leadman in the cannery talked a supervisor into hiring a worker, or an opening was created when a worker was fired, the lead filled out the requirement on a card and handed it to one of the assistant supervisors.

He hustled out to the bull pen; people rushed to the gate. The supervisor called out the type of jobs, how many men or women, hourly wage. Those who could handle it or wanted it badly enough, picked up cards and moved through the gate.

We became friendly with an Arkansas family. They told us that workers were needed. It was the height of the peach season. They said that if we could just get inside the cannery, we could get a job. So I devised a plan..

One of the family members was an attractive girl our age. I suggested to the girl that she go over to the gate guard, "sweet-talk" to him, get his back turned away from the open gate and keep him distracted while we ran across the yard to the cannery. She could then talk the guard into letting her in. She agreed, and did it.

My sister and I ran to the building and raced up the stairs to the office where a confused clerk signed us on for the night shift. We were walking down the stairs with our time cards as our smiling friend came in.

Our leadlady told us to put the time cards in our shirt pockets. She would punch them with a little hand punch as we started work.

She put us to work canning halved peaches. We pushed them into the holes containing the cans. At the same time I tried to prevent slugs from going into the cans, too. This slowed me a little and did not make me popular with the leadlady. Working in tomatoes was worse, a new acquaintance told us. He was a "worm parts counter". The catsup contained everything, but there was a maximum number of parts allowed, you see. I guess if you can think of them as protein.. To this day I have trouble eating canned food.

The worn-out machinery was continually breaking down. Those with jobs immediately dependent on that machine had their time cards punched out. They remained at their stations while the mechanics fixed the machine. As the work stopped down the line, more cards were punched out. When the machines started, each card in turn was punched in. We spent more time punched out than in. Our shifts were for 12 hours at night, but we might have six paid hours if we were lucky; more often it was three or four.

In August 1941, Betty and I cashed that week's paychecks of $13 each and took an earlier bus to Fullerton Airport for our first 30-minute flying lesson.

Fullerton Airport was a grass field with hangers and offices at the roadside and orange groves nearby. The air was clear and fresh and a light afternoon breeze tempered the summer heat.

Several male flight instructors hung around the offices, looking for students. They charged $2 an hour, a huge amount to us, and as I remember, the planes cost $5.

We usually bought 30 minutes each time. We would have one instructor for a time or two, then he would disappear.

Some of the instructors we flew with were fairly good; most were impossible. One was too lenient; the student learned poor habits. One idiot was scared of his students. He planted his heavy feet on the rudder bars, gripped the control stick in deathlike stiffness, and yelled at the hapless student.

After one lesson we avoided him. We worked too hard for our money to have him waste it.

Sometime during our high school days, our family rented a home built at the turn of the 20th century, on a street of like homes across the street from the two-year Santa Ana College.

Our home had huge sliding doors between the big living and dining rooms, and a number of bedrooms, some of which Mom rented out.

While I was still living there, a movie company filming "Gallant Journey" set up their equipment in front of a beautiful warm cream-yellow home next door.

This home portrayed the residence of the fiancée of John J. Montgomery, the San Diego glider builder who in 1883 made the world's first controlled winged flight in a glider. Glenn Ford played Montgomery. In a sequence where Ford rushes Janet Leigh out of the house onto a bicycle and pedals down the street, our home is seen for one split second.

In September, 1941, we enrolled in the two-year college across the street, and when CPT (Civilian Pilot Training) ground school began on the campus, we were allowed to attend it "because we were already flying," as the registrar put it. The only girls in a room full of handsome young men.

I planned to leave for flight school when I had enough money after graduation and that school might not have formal ground instruction. Even if it did, the more I learned, the safer I would be. I planned to fly for a long time.

Japan bombed Hawaii on December 7, 1941, and the federal government limited all private flying to 150 miles or more from any coast. Our last lesson at Fullerton was November 16, 1941. Fullerton is about 12 miles from the west coast.

I asked the Civil Aeronautics Authority for a list of "approved schools," and sent for brochures from flying schools in desert areas in California, Arizona and other inland places.

From the costs quoted, I needed about $350 to live and fly at one of the schools until I had 35 hours of flying time. Then I could work at something until I was accepted by the WASPs (Women's Airforce Service Pilots), formed in 1942.

Flying as a WASP meant flight time in many kinds of aircraft, helping the war effort while moving my plans forward.

Men and women ferried planes from the aircraft factories to Canada for others to deliver to England. Some British women flew the larger planes on to England. In England, women ferried planes as members of the British Air Transport Auxiliary. In September 1942, America's experienced women fliers formed the WAFS (Women's Auxiliary Ferrying Squadron of the U.S. Army Air Forces Ferrying Division). Later it was absorbed by the WASPs, headed by Jacqueline Cochran.

Age minimum was 21 and I was 19. Later the age was lowered to 18½. Thirty-five flight hours logged was required before applying. Women were sent to Sweetwater, Texas, to finish training "the Army Air Force way."

More than a thousand women fliers became WASPs. The ex-WASPs I know are remarkable women, and many are responsible for the opportunities that today's young women have in military air service, as airline pilots, and in general aviation careers at all levels.

The canning season ended and Betty and I worked for the next two years in a drive-in restaurant. I banked every cent above college costs and other expenses from my $18 a week. We lived at home and Mom never charged us for anything. She was a hospital nurse's aide and rented rooms in our home to make ends meet.

Betty and I graduated from the two-year college in June 1943 and thought it might be fun to get paid jobs in a summer camp. I had over $350 saved and planned to leave for a desert flight school in the fall when the weather cooled.

The San Francisco Boys Club camp in the coastal redwoods above San Francisco needed help and would operate for two months, an ad on the college bulletin board said. Among the jobs was one for a nature director and another for a lifeguard. Each paid $25 a week plus room and board. I got the job of nature director and my sister, the lifeguard job.

I borrowed library books on coastal redwood ecology as I knew nothing about the subject. My twin had her lifeguard card.

At the camp were two women, a nurse and a counselor for the younger boys. The men counselors were draft deferred students of medicine or the ministry or those who could not pass military induction tests due to physical disability or age.

Nature director duties included caring for the small museum and conducting nature study hikes. Caring for the museum meant sweeping it out, dusting exhibits, and keeping all in repair.

That first day, the young campers had caged in their tents numerous harmless baby snakes. The boys arrived at the museum before me the next morning, put a bunch of snakes inside the museum, tied several around the door knob, then hid.

I eyed the squirming snakes on the door knob and heard the giggling behind me. I pasted on a smile, gently unwound the little snakes, and placed each one carefully on the ground. I was saying loudly how "cute" these were as the grinning boys came out of the bushes. In the museum, snakes were everywhere, including some slithering down the staircase. But, there were no more tricks and we had some great times together.

Camp closed at the end of one month; polio had hit the camp, tragically. After our quarantine lifted, we returned home to Southern California. A letter was there from a flying school at Conner Field, Quartzsite, Arizona, inviting me to come.

Chapter 2

The interstate bus left me on the edge of the highway in Quartzsite, a place that was only a blister on the road. The dust settled as I picked up my belongings.

Across the highway, a narrow road led to wooden barracks-like buildings near a huge white square bulldozed and leveled from the desert.

I crossed the highway and walked down the road, face and arms reddening from the hot sun. Beyond the building marked "Office" was a long doorless hanger close to a taxiway with two-seater 65 horsepower Piper Cubs near it. One fuel pump was in front of a shed filled with oil barrels.

I put down my suitcase and bedroll in the office. A whirring fan sent a welcome breeze through the room. School owners Johnnie and Bessie sat in back of a counter partially covered by manuals, magazines, and flight charts.

They greeted me warmly. I filled out the registration form and gave them a check for 35 flight time hours, dual instruction and solo time, the WASP acceptance minimum. The fee also included a bed and a locker but no meals.

Bessie took me out to the women's dormitory, showed me my bed and a locker and introduced me to the dozen women resting, writing letters, or studying flight manuals. Most of them were my age as the WASP minimum recently had been lowered to $18\frac{1}{2}$. The girls were serious, intense, accepting the heat and dusty discomforts for a chance to join the WASPs.

Water for the airport came from a tank on top of a nearby hill. Pipe lines laying on the ground, ran down to the buildings. The sun heated the water in the pipes for the shower rooms.

I found that taking a shower had to be planned well. During hot weather, one must shower late at night when the water had cooled down. In the winter, we'd shower during the day when the water was warm. Before one showered, though, it was prudent to sweep out the black widow spiders, tarantulas, and scorpions.

I walked outside when the sun was low. The dry air had cooled and the hills, brought near by the clear evening air, had shadowed into light amethyst. Cactus, sagebrush, and mesquite dotted the landscape. Piles of sand and rocks rose among them.

"Come on, Claire," called someone from a knot of women moving out of the barracks. "We're going to have dinner at Elmer's."

Men from their barracks and others from the line of planes joined us. Daylight darkened into velvet twilight and the first stars appeared.

The next morning I had my first lesson with my instructor, Ralph Grill. We checked the plane over together, untying the safety lines from the wings as we walked around it. Ralph told me the plane's brakes did not work so we would leave the wheel chocks in place until after we started the engine. This was a Piper Cub J3, a high-wing, tandem (one seat behind the other) with a tail wheel. Nearly all planes had tail wheels then, including the famous Douglas DC-3.

I climbed into the rear seat from the right side of the fuselage. Since a solo pilot sat in the rear to help balance the plane correctly, a student learned to fly from the rear seat.

The three gauges — compass, airspeed indicator and altimeter — were mounted on the instrument panel in front of the front seat. The fuel gauge was a straight wire with a cork on the end floating on the fuel in the tank mounted directly in front of the cockpit windshield. The top end of the wire was bent so it would hang on the tank top when the fuel was gone. The tank held 12 gallons — three hours flight time.

I moved the control stick and the rudder pedals back and forth lightly; all moved freely. In the cloudless cool dawn, pink light spread over the desert, followed swiftly by a golden light. Too soon it faded to white.

After soloing, I solved the no-brakes problem on this Cub by turning to one magneto instead of two while taxiing after landing. This slowed it enough to control its forward movement to its parking space. One of the other students tried this, except she tried it with a one-mag plane and had to push it in from the runway.

Ralph stepped in front of the plane and reached up for the propeller blade. He pulled it down easily, then pulled it down once again. He yelled, "Switch on. Throttle cracked." I yelled back his words, turned the magneto switch to "Both" and pushed forward on the throttle just a hair.

Ralph pulled the prop down hard through its arc and jumped back. The engine coughed; I worked the throttle and it caught. Ralph kicked the chocks out of the way and climbed in. We taxied out to the take off line while the engine warmed, checked the controls and magnetos, looked in all directions for any other traffic, saw none, and took off.

Seven flight lessons later, Ralph said I would solo with one more hour of take offs and landings. He had to go to Los Angeles for the weekend; another instructor would fly with me.

This instructor was hired just a few days before, but already his students hated him. He spoke to a student only to give clipped

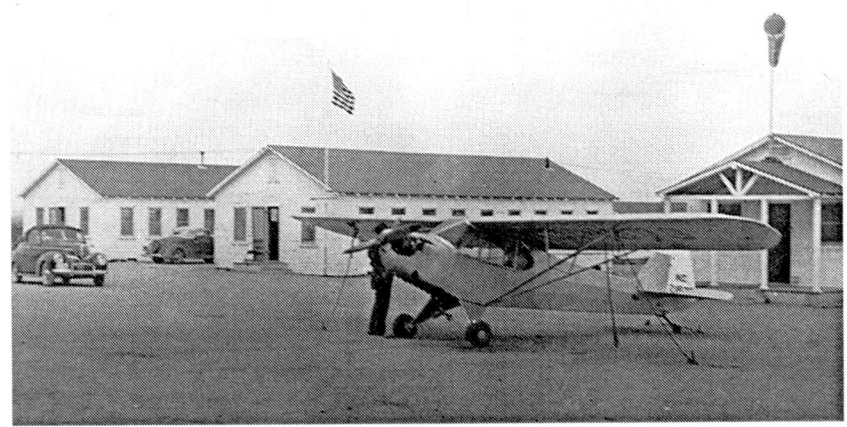

**The office and barracks at Conner Field
Quartzsite, Arizona, 1943-44**
Photo courtesy of Dixie C. Leerskov

orders, spent no time on the ground discussing the hour's flight lesson coming up, and shouted instructions over engine noise without turning his head to look at the student.

We taxied to the tie-down space after an hour's confusing flight. The instructor told me I did not know how to fly and would never learn. He swaggered off. I sat stunned.

Confused and angry, I went to the school owners. "He said I would never learn to fly. Ralph told me I was ready to solo. I want another instructor or the rest of my money back."

Of course they couldn't give me my money back — like most schools, they had already spent it. Steve Edwards, chief pilot, was asked to give me a lesson. We went out for 45 minutes of stalls, spins and landings. Then he soloed me. I shot three beautiful landings and felt redeemed. It was October 10, 1943.

The other girls watched my solo and ran to the plane, grinning and yelling "She's too hot a pilot, throw her in the shower!" We had no swimming pool.

The rotten instructor? He was fired. A student pilot with 78 hours, he had forged his instructor certificate.

After ten solo hours practicing straight and level flight, climbing and gliding turns, 180s and 360s (half-circle and full circle turns), many take offs and landings, stalls and spins, my instructor and I went on the required dual cross-country to Phoenix, Wickenburg and return.

I was now ready for my solo cross-country. I scheduled a Cub for an early morning flight, but the school kept moving the schedule. It was afternoon before I took off. Landing at Phoenix Airport, I refueled the plane for the return flight.

A mechanic helping me said it was getting a little late; it might be better not to go. I looked at my wristwatch. "Oh, I think I can get back before dark. If not, I'll land someplace along the way, call the school and go on to Quartzsite in the morning." Years later I found a similar statement on a list of famous last words.

Reluctantly, the mechanic propped the plane. Climbing to my flight plan altitude, I trimmed the plane for cruising and checked the chart against the tangle of roads beneath me to be sure I picked up the right one to follow. Chart, compass, sun position, and landmarks all checked. In a little while I looked at my watch to record the time on the flight card... a chill crawled over me. My watch was stopped.

The ground was deep in shadow, the sun low, but it was light at my altitude and the evening air was smooth. I looked for an airport, but there were only emergency strips without telephones or buildings; I'd have to stay in the plane all night. Then it was too dark to land at an unlighted, unfamiliar air strip.

The instruments on the panel beyond the front seat became too dim to read. No one flew at night from our unlighted field. My plane had no working lights and no flashlight; we were invisible.

None of our training planes had radios. Airport tower operators used red and green light guns to direct landings and take offs at their fields for those without radios.

Mountains lay between me and home. Was I high enough? Frightened, I went higher; panic gripped me, tying my stomach in knots. I calmed and started thinking about the problem. I looked down and saw car headlights shining. A rotating red and white beacon on an airway flashed some distance ahead, right on my route. Quartzsite Airport was on that airway. All I had to do was stay in the air.

In those days, airways were marked with red and white rotating beacons. Large airports had green and white rotating beacons, and still do.

A cluster of lights appeared: the lights of Quartzsite, then the school building lights. The field was a white patch in the night. The field was 900 feet above sea level, but my altimeter was not readable in the dark.

I mutter to myself as I plan my landing, "If I bring it down gradually and straight in — there won't be any other traffic, I hope — the plane will bounce when the wheels hit, and I'll work it back down with a little power—easy now."

The Cub touched gently, bouncing only a little. I taxied to the line, turned off the switch, and sat there, knees shaking, catching my breath, wiping my sweating hands on my slacks.

Everyone was outside, having heard the engine. They raced to me. Johnnie and Bessie angrily scolded me for flying at night but the students cheered.

Several of us finished our 35 hours and sent in our WASP applications. That week the WASPs announced that their ranks were full; they were not accepting applications for awhile.

My money was spent. The school hired me to work in the office. The noise bothered me and I asked to work in the hanger patching holes in the fabric on the airplanes. I flew an hour each day during the week and attended ground school.

The WASP aspirants departed and new students arrived. Most of these were learning to fly because it was the fad that season. It was also an excuse to run away from parents or husbands.

One of the women students starting at the school the same time as I, was Dixie C. Leerskov of Bakersfield, California. She sent in her WASP application the same week I did. She went to work in the office; I, in the hanger. Hers was the better deal, I think, but I liked working in the relative quiet.

I worked in an open hanger where the endless windy days whipped in sand. Burrell, the mechanic I worked with, was a big, burly, foul-talking fellow. He spent most of his time on the engines, keeping them going in spite of the sand and the students. He had drifted in one day and a few months later he drifted away.

Years later I saw Dixie in Bakersfield where I spoke to the Bakersfield Chapter of The 99s. We talked of those times and in a letter to my sister, she wrote also of the fun we had had.

The big bomber base at Blythe had handsome cadets and lieutenants who piloted B-24s. They frequented Elmer's Café/Bar near our field and the Hecklehut Café a little further away.

It was exciting. Pilots in trainers from Phoenix stopped at the field. Once a P-47, my dream plane, landed. I spent some time close to it, dreaming of the time when I'd be in the WASPs and flying one. But that was not to be.

About 200 Civil Air Patrol members came from Los Angeles on the weekends. They took over the planes and while we could not fly when they were there, we did have fun. Hollywood people such as Alvino Rey, with whom I flew, and Ray Milland, who shared my cream puff one time, added a special glamour to our lives.

Hot sun and dry air weakened fabric and plywood on the planes. Students grinding through brush at the sides of the runway and the wind blowing tumbleweeds and small rocks against the fuselages punched holes in the fabric. I patched them. When fabric could no longer pass the aircraft inspector's tests (the sharp pencil he dropped on a wing went through it instead of bouncing), the mechanic and I took that plane apart.

We placed the wings and tail surfaces on wooden horses, and put the fuselage up against the back wall for the time being. After I stripped off the fabric, the mechanic checked the condition of the frame. Then I laid new cloth on loosely and glued it down all around and stretched it tight by sprinkling it with water. I brushed successive coats of clear nitrate dope evenly on both sides of each surface, following that with silvered dope; the final coat was the Piper Cub yellow.

It was hard, mean work, done in an open hanger. Afternoon winds blew sand on my work, my clothes, my face. Dope got on my hands and sometimes caused my skin to crack.

I did a careful job and was proud of my work. Federal CAA (Civil Aeronautic Authority) aircraft inspectors checked my work each month. They always approved it, always complimented me.

Repairing these planes paid my living expenses and flying time until I received my commercial license in September 1944, one year after I came to the school.

Flight Inspector C.A. "Al" LeFevre flew with me on my private license ride after observing from the ground my spins over the field and my spot landings. I climbed the Piper Cub J3 to 3000 feet above the airport, spun three turns to the left and three turns to the right, then made three spot landings. After that Inspector LeFevre climbed into the airplane for the rest of the ride. He gave me my private license in March 1944, long after I was eligible. I wanted to know I could handle any problem as someone would surely ask to go with me. I have always felt responsible for the safety of others in my planes.

Once a month two CAA inspectors flew in to Quartzsite from Phoenix, then on to other airports, and back to Phoenix. One was a flight inspector and the other, an aircraft inspector. These men "rode the circuit" of airports. The CAA is currently the FAA (Federal Aviation Administration).

The aircraft inspector pressured me for a date: "I understand you are taking your ride today. Are you going out with me tonight?"

"I don't think so, I don't plan to."

We went through this routine several times; finally, I agreed to go with him. I made sure, though, that I stayed with a group.

Nearly all of us got our private licenses that day and our spirits soared. On to the dance. We loaded into cars and on the way stopped at Elmer's for a beer. It was a favorite place for pilots.

My current boyfriend, a lieutenant at Blythe Air Force Base, sat at the counter. Earlier that day he broke our date because he was going to the dentist. He was angered to find me with the aircraft inspector. I was not sympathetic. We were only friends, and he had not bothered to call again.

The young man stamped off. A couple of days later he sent a fellow pilot as his emissary to discuss the matter. I thought this was funny. There was nothing to discuss; he need only ask for another date. And he did, eventually.

For one brief moment in 1944, the WASP program opened again and we went into Los Angeles to a hotel where Ethel Sheehy, executive assistant to Jacqueline Cochran for the WASP Program, as well as president of The Ninety-Nines, international organization of women pilots, interviewed applicants.

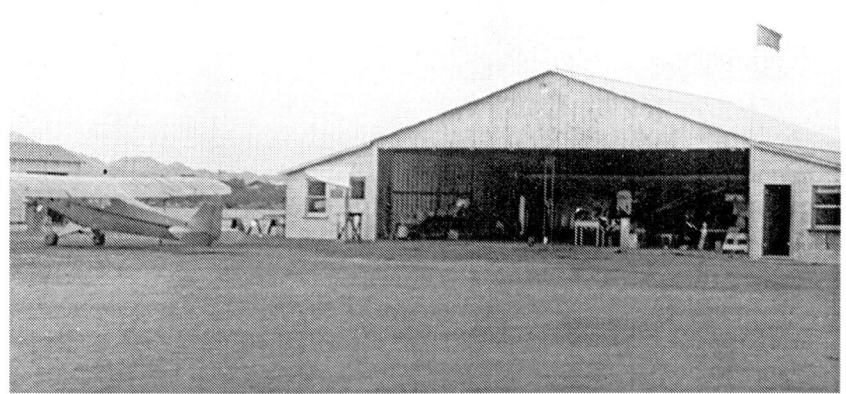

Open hanger at Conner Field, Quartzsite, Arizona, 1943-44 where Claire repaired fabric on the Cubs.

Photo courtesy of Dixie C. Leerskov

Mrs. Sheehy looked at my private license and asked me to join The Ninety-Nines. She signed the membership application as my sponsor. I have always been proud of that fact. I was elated. I had heard of that most elite organization of women pilots when I saw a picture of Betty Gillies on the cover of a magazine some years before when she was president of The 99s.

Those of us who passed the interview took Army Air Corps written tests. As we drove back to the desert the radio said that the WASP was to be deactivated. A week later a letter came saying I was accepted for the flight school if the WASP was reactivated. It never was.

One of the fun things I do today is join in meetings and fly-ins of The 99s. In 1971, I was Governor of the Southwest Section and spoke to many of the Chapters. The 99s' first elected president was Amelia Earhart. Jacqueline Cochran was president in 1941-43, just prior to Mrs. Sheehy's term.

Today it is a vital group of 6000 women engaged in every aspect of aviation: sports pilots, flight instructors, crop dusters, flight school owners, military pilots, airline captains and flight officers, and astronauts. Many members are nurses, teachers, homemakers, flying for the fun of it. Those no longer flying contribute knowledge and time to many worthwhile aviation activities.

Membership often leads to interesting encounters. Some years ago my friend Elizabeth Dinan, a fine flight instructor, race pilot, and a 99, was on an Alaskan bus tour with her husband Don. A bearded man in rough clothes stepped out to the roadside from the forest. He got on the bus and sat near my friends.

Noticing Elizabeth's 99s pin, he said that many years ago (had to be around 1950) he had taken some flight instruction at Bellflower Airport from a tall blond woman named Claire who belonged to The 99s.

In June 1944, the flight school moved to Manzanar Airport, California. Johnnie and Bessie told us about it one evening and said three of us were to start the ferry flight at dawn, each flying a 65 horsepower Piper Cub trainer. Flight instructor Fred Brach was to fly one, Ellsworth Gibbs, a mechanic and private pilot, another, and I was to fly the third.

Brach and I were to take students while Gibbs wanted to fly alone. Ethel Majors was to be my passenger.

The decision to move was so quickly made and carried out that we had no time to plan our trip well. We searched the files and found one two-year-old sectional chart of the area we would fly over. Brach was leader; he took the chart. My cross-country time totaled four hours dual and eight hours solo.

At dawn we warmed the engines and discussed the flight again. It was about 350 miles straight across the desert. Silver Lake Airport at Baker, California, was our first stop; Inyokern, our second, and Manzanar, our last.

Brach removed the control stick from the front seat section on my plane as Ethel was a student and I was not an instructor. This was a CAA regulation at that time. Brach got the weather report and filed our flight plans to Silver Lake by telephone.

Just before take off we checked the windsock and the sky. The sock was limp, the sky was clear. Brach took off first, I followed him, and Gibbs was last. Our planes had no radios. Soon after our take off, we encountered a rising headwind and some turbulence. An hour out our time between check points was almost 100 percent more than we had estimated. I watched the lead plane, silently begging Brach to turn back.

My cross-country flying instruction had not included fuel conservation, nor could I adjust the mixture control. My plane was slower than the lead Cub and I used a higher power setting to keep up, using even more fuel per mile flown. Cruising speed was about 60 to 65 miles per hour and with a 30 to 35 miles per hour headwind, we were doing poorly, to say the least. The air got rougher as the sun heated the land.

The fuel gauge wire slipped steadily into the tank and I knew Silver Lake was out of reach. I stayed on course directly behind Brach and kept him in sight until he was just a speck. Then he vanished. Beneath us were gullies, boulders and steep hills. We flew on, looking for a place to land.

Frantically, Ethel and I looked for Gibbs, supposedly behind us. If he saw us go down, he could send a search party.

There was no plane. We were shocked. Had he gone down? We had no fuel — we could not go back and search for him. The bent end of the wire fuel gauge hung on the tank top. We saw no landing place and I felt sick. Then we flew over Granite Peak and out of the rocky region into flat desert country. Railroad tracks appeared straight ahead.

I said to Ethel, "It looks like we are going to have to land near the tracks and take the next train to town!" The engine sputtered and died.

My sick feeling vanished. I concentrated entirely on the landing. We were about 3500 feet above the ground. This altitude gave me time to set up my landing pattern. I glided over the railroad tracks and watched for dust devils to give me the ground wind direction, as there was no smoke. I must land straight into the wind which seemed strong enough to equal our stall speed, lessening our danger. I glided over the one-lane dirt road that twisted and turned towards the station. A clear spot appeared near a little station (named Glasgow, I later learned). I stalled the Cub ("landed" it) about three feet above the sand, preferring to risk a damaged landing gear to having the plane flip over. It rolled about 40 feet before the wheels sank into the sand.

Ethel and I unbuckled our seatbelts and wearily climbed out. We inspected the plane for damage and found a few small rips from the sagebrush in the fabric on the elevators.

We walked to the station where men worked on the roadbed of the tracks. They had not heard our plane because of the wind and were startled to see two tall young blonde women appear in their midst. They gave us five gallons of gas, carrying it to the plane for us.

For the next hour we searched for a take off place. The only bit of straight and level hard ground was the gravel bed between the double railroad tracks, the siding section.

We measured carefully. There was room for Cub wheels between the two sets of tracks, with inches to spare on each side. The telephone wires on poles at one side of the tracks were far enough from the tracks to pose no real danger if I kept the plane straight ahead on take off.

Signal devices were several hundred feet down the tracks. Right here was my runway.

The job foreman worried the plane would crash and burn. He asked me to get a tractor from somewhere to tow the plane to a paved road at Kelso. I thought only of time and expense involved and refused. He was right, of course; that is what I should have done, or something like it, and I'd do that today.

I planned the take off in detail. The run must be short because of the obstructions and a slight cross-wind. The wind was blowing 35 to 40 miles per hour. I could not let the plane drift because of the wires. Now the wind was my ally in this short field take off just as it was in landing.

Trains ran 20 minutes apart. The men pushed the plane close to the tracks. The next train passed and the men lifted the plane and placed it between the double set of tracks.

I stepped into the plane sure of myself, but as I fastened the safety belt, butterflies started fluttering in my stomach.

What kind of a fool was I? My silly pride kept me from backing out. Ethel was sure I could do it and refused to remain behind.

Since the plane had no brakes, I asked four of the husky men to hold the airplane by the horizontal stabilizer and the wing struts. Ethel pulled the propeller through and the engine roared. She climbed in. I moved the throttle steadily forward until the engine was at full power. Ethel yelled for the men to turn loose.

The plane jumped forward and rolled down the "runway." The right wheel struck a rail. Fearing a crack-up, I eased back on the stick, bringing the nose up. The plane lifted from the ground. I put the nose down, picked up flying speed, and the plane climbed above the wires.

After gaining a few hundred feet, we flew back over the workers and moved the Cub wings in a goodby-and-thank-you. We turned and flew on to Silver Lake, 20 miles ahead. A few minutes later we spotted Fred Brach's plane and when he saw us he looped his ship, then led us to Silver Lake. Brach and Gibbs met us near the runway, steadying the wings as we taxied slowly to a parking space.

Ethel and I crawled out shakily and helped tie down our ship. The wind was still about 40 miles an hour. Brach told us he had put a five-gallon can of gas in the front seat of his plane and retraced his flight, figuring that I'd be right on course.

Gibbs had reached a side road near Kelso and got gas. He had left us just before we went down. Brach was able to conserve his fuel, landing at Silver Lake as the tank ran dry. He and his passenger pushed the Cub off the landing area.

We refueled and held a meeting. The wind was strong and gusting, ripping other aircraft loose and flipping them. It was safer to continue the flight than to stay at Silver Lake.

Brach gave the chart to Gibbs; he was to lead us to Inyokern. After flying two and a half hours we reached Highway 395, but no airport was in sight. Gibbs circled over the highway.

After watching for a few minutes, I landed on a little dirt road near the main highway. My fuel was low; I dared not waste it while Gibbs decided on a direction. I'd flag down a motorist and ask him. The other planes followed me in.

The chart showed we were 20 miles south of Inyokern. A passing motorist told us Inyokern was, indeed, 20 miles north; it was the first airport we'd see, and it was right next to the highway.

Ethel propped our plane and we took off to follow Highway 395 north. The other planes followed. We looked for the small field indicated on our chart and found long paved runways of a military field.

Entering the traffic pattern, we watched the tower for a signal. No red or green signal light flashed. Our fuel was too low to wait; we landed.

A young sailor met my plane at the taxi line and told us we were on a Navy base. He said I should have answered the radio call from the tower. I told him our plane carried none, offering to let him search.

The other planes landed. The Navy graciously gave us fuel and offered any assistance we needed, such as weather information. They were nice in view of the fact that we were on a military field without permission during a war. Inyokern is today an Inyo County airport.

We took off north for Manzanar Airport. The wind was still blowing and the sun was low. We flew over Lone Pine and saw Mount Whitney on our left, the highest peak in the Sierra Nevada. We were almost to our new base. Manzanar Airport came into view, across the highway from the Japanese-American internment camp. Independence, population 600, was one mile north of the airport.

Owens Valley is desert, with the great Sierra Nevada range on the west and the White Mountains on the east. The high mountains cut off the lowering sun, casting a deep shadow.

We landed in twilight. A mechanic met us near the taxi strip and held the wing steady as we taxied slowly to a tie-down space.

Ethel and I climbed out, so tired we could hardly move. We had taken eight and a half hours of flight time to go 327 miles, averaging 37 miles per hour. We made so many mistakes. One old chart in another plane; none in ours. We did not have good weather information, continuing our flight in heavy headwinds.

Flying a 65 horsepower plane straight across the desert instead of following a highway. One forced landing in the desert because I lacked experience in conserving fuel, and then using space between two sets of railroad tracks to take off. We could have cracked up. I'll never do that again. The only thing we all did correctly was to file the flight plan. And that was required by the federal government.

Chapter 3

I had fun in Owens Valley. One of our school's visitors was a very young Bill Lear, Junior, who asked me to take him up in a Piper Cub and do some spins. Years later, in his late teens, he visited me in my Santa Ana home and offered me a ride in his P-38. He was performing with it in air shows. Unfortunately, I could not accept: I was wearing a heavy brace for the back injury I received in a street-car accident. He played our piano for me.

Another fond memory involves my Marine Corps pilot friend — a captain stationed at Mojave. He'd fly his Corsair above Manzanar. I'd join him in a Piper Cub and fly formation — me at full throttle, he with reduced power, flaps and gear down. What fun! At night he'd drive up and we'd go to Keough's Hot Springs at Bishop for dancing. He was a good friend. Then his group was gone.

Two weeks after I received my commercial license in September, 1944, I was ferrying airplanes for Les Farrar of San Bernardino. He bought and sold new and used aircraft. I had 216 flying hours logged; 200 is the minimum for a commercial rating.

I had met Les somewhere. He told me that when I got my commercial, he would give me a job ferrying airplanes. I did not know if he was serious or even believed I would get my commercial rating, but I called him. "I'm ready."

"Come on up."

Les Farrar's brother was vice-principal (later the principal) of Santa Ana High School when we were students there. Robert Farrar was scholarly; Les was a businessman.

Les bought lightplanes such as Piper Cubs, Luscombes, and Taylorcrafts, from around the United States. These had been used to train men going into the military air services. Each student received ten hours of flight instruction before he was sent to a training base. This system weeded out those who realized they did not like to fly as well as those who failed to learn.

As World War II wore on, these old planes were replaced by bigger trainers with more horsepower. Les bought the small planes from ads in the aviation newspapers. He never saw them, just bought them over the telephone. Les employed a number of pilots to fly them back to Southern California. We could fly to the California coast and out of it, but not locally. There was no danger from enemy invaders; rather, it was from our own nervous young antiaircraft gunners.

Many of the 60 to 85 horsepower engines on these planes were in rough shape and barely ran. Too often the fabric covering was

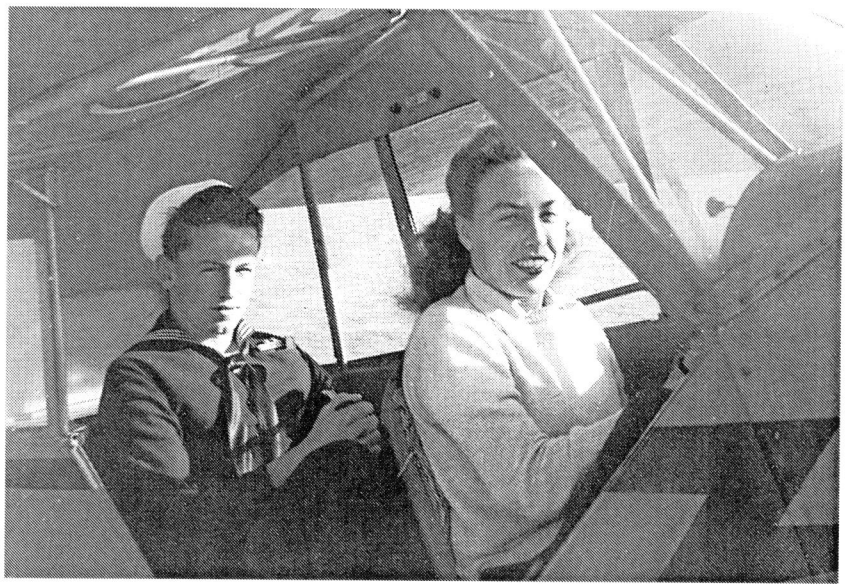

Deane McMillen, our brother, with Claire in a Piper Cub. Thousands of beginning students in the 1930s and 1940s first took to the air in the Cubs.

Photo by Betty Loufek

also in bad condition. Overworked, abused by endless numbers of students, repaired by minimally-trained mechanics who lacked the necessary replacement parts, the planes now sat rusting, fabric rotting from the sun.

My experience at Quartzsite in recovering and patching fabric helped me. The fabric often was so rotten I could drop a pencil vertically on the top of the wing and it'd go through. Fabric failure caused many accidents when others flew planes in aerobatic maneuvers or in turbulence. The cloth simply shredded.

Once after refueling a plane at Blythe, California, I stood up under a wing and put my head through it. I used adhesive tape from my first-aid kit to make a temporary repair.

Before each ferry flight I checked fabric, bolts, safety pins, rigging, controls, instruments, tires, brakes, and all visible spaces for trash such as oily rags, metal parts, tools, and paper. I studied the aircraft maintenance records carried in the plane. These often told me of problems.

I made my flights close to a highway. Many times I wore a parachute but, happily, I never had to use one.

This habit of checking everything has stayed with me through the years. I do it whether it is a plane or a car. Its importance for airline flight officers was underscored a few years ago when a mechanic removed close to 50 bolts from the tail section of an airliner and no one made sure they were replaced. This part fell off in the air and all aboard died.

Frequently I am an airline passenger and though I cannot inspect, flight officers can but often don't.

One of the places where I'd refuel was Silver Lake Airport, a marked off mile square on a dry lake near Baker, California.

I'd land, refuel, then taxi across the airport to the CAA radio station. I'd check on the weather and file a flight plan to Fullerton or to San Bernardino Airport.

There was a corridor that we had to stay in while flying to the coast. This was 1944-1945.

I always looked forward to visiting with the girls that ran the station. Working there were Joyce Carl, a blonde, and a tall redhead, Jean Larson. Both were in their late teens. I was 20. Both became well-known in aviation circles. You know them as Joyce Failing and Jean Schiffmann. We have been friends now, as of 1999, for 55 years.

I ferried the small planes from Salt Lake City to Fullerton Airport, from Dallas, from Idaho Falls. Name a town — I've probably been there.

One time I flew from Idaho Falls to Fullerton and called Les.

"I'm in."

"Fine. I just sold it to someone in Clovis, New Mexico."

He bought it and he sold it over the telephone. I delivered it to Clovis, took care of all the paper work, went to the bank with the man for the check, closed out the deal. This was tremendous business experience for a young woman who planned to open her own flight school.

I celebrated my 21st birthday while ferrying an airplane from Georgia. Traveling by bus and train for this, I started right back even though I was tired. Once again there was no place to stay overnight where I picked up these planes. Going to Georgia took four days and nights. I picked up a Taylorcraft in Griffin. This was my first flight in one so I had an instructor fly with me around the field. I flew it to the Columbus, Georgia, municipal airport for weather information and flight service. I was exhausted and in danger of falling asleep at the controls.

Parking the plane for the night, I went to a hotel. The clerk said he had no room. This was always a problem. Women traveling alone had difficulty finding a decent hotel room.

Clerk attitudes did not change until in the 1980s when hotel managers learned that 50% of business travelers in the United States were women, and this percentage was expected to rise. In the 1940s, clerks were suspicious about our intentions.

Sometimes the hotel was nearly full and they did not want to rent to one person when they could soon charge for two. But I believed this clerk thought I was a prostitute, perhaps cutting in on his own action? I demanded to see the manager.

I introduced myself to the manager and said, "I am ferrying an airplane. I'm tired. I've come here to your hotel to get a room. The clerk tells me there aren't any, but I have a feeling there is and I want one. And I want one now, and I want to sleep." I got a room.

Next day, the weather was too poor for flying. The hotel manager drove me around the city and across the river to Phoenix City, Alabama, showing me various points of interest. He was nice and really looked after me.

I soon learned that hotel detectives could not be trusted. Especially in three of our 50 states, men thought they were the world's greatest lovers: Texas, Oklahoma, and Arkansas. Not just hotel detectives, but other hotel and airport personnel. Many were too ready to "protect" an unchaperoned young female and thought it could best be done by sharing the same room. I put a stop to such nonsense for myself.

Sometimes only two-person rooms were available because of a wartime regulation in a town. Then I phoned for a reservation for "Mr. and Mrs. Claire McMillen." At the hotel desk I signed in and said that my husband would be in shortly.

Ferrying these airplanes did not pay much, $25 and expenses for each trip, I think, but it was fun. Handling different aircraft was good experience, as was navigating many routes in varying weather conditions and dealing with people.

At times I had sore knees because some planes were not built for legs as long as mine. I am five foot seven inches tall; I wonder what size person the builder had in mind?

Sectional charts cost money so I used the map of the United States on the back of a California road map. Airports were marked with red airplane figures. Roads followed passes through the mountains, and generally, I followed roads. I dared not be far from a road at any time.

Airports were not close enough together for these short-ranged airplanes. Often I took off knowing the plane lacked enough fuel to reach the next airport. A Piper Cub J3, for example, carries 12 gallons, using four gallons an hour for three hours, at 60 to 65 miles per hour.

Not enough range for one airport to another, and westerly headwinds made it even shorter. I planned for these off-airport landings.

Carrying a five-gallon can of fuel strapped in the second seat, I flew until the tank was nearly empty. Then I'd find a big farm, pick out a nearby road, circle to check wind direction, fences and other obstructions, and to attract the farmers. Five minutes after landing I'd be surrounded by people wondering about my 'forced' landing. I asked the largest farm boy to pour the fuel into the tank; showed him how to "prop" (start) the airplane by pulling the propeller through a half turn. The trainers I ferried had no electrical starters. Flying to the next airport usually took another hour and a half.

Sometimes I flew to a landing strip where pilots buzzed the service station on the road, landed and waited for the manager to send his fuel truck. Landing on this strip once, I taxied down the dirt road, onto the highway and up to the service station pump. After the fill-up, I taxied from the pump and took off on the highway itself as there were no cars in sight.

In Salt Lake City I picked up an open cockpit Fairchild PT-19 with a 175 horsepower Ranger engine. After checking out in it between snowstorms, I started back to Los Angeles. I wore a parachute as well as a helmet and goggles. I was feeling good about flying this larger airplane and wearing these all-time pilot symbols. I refueled the plane at St. George, Utah.

Carrying helmet and goggles with me I swaggered into a café for lunch. The waitress looked out the window and said in a puzzled voice, "Well, gal, where'd you park your motorcycle?"

Several men and I went to Blythe, California, to ferry Vultee BT-13s. Each plane had a 450 horsepower Pratt and Whitney engine. A checkout pilot had me taxi for 30 minutes, then we took off and circled the field once as he showed me how to fly it. He brought it in and bounced it high on landing. I flew it around the field and made a mistake, a perfect landing. He insisted I take it to the west coast without further practice. I did it even though my knees shook so I could barely keep my feet on the rudder pedals. I spent several minutes circling Orange County Airport (John Wayne) before I had enough courage to land.

Ferrying a brand-new Piper Cub from the East Coast, I was nearing Indianapolis when thunderclouds appeared, pushed by 70 mile an hour winds. My plane could not fly fast enough to beat the storm. Landing, I taxied fast to the hanger, but its doors were closed against the wind.

Two men grabbed the wings to tie them down. A wind gust hit, lifting the plane off the ground. The men turned loose and ran out of the way. Power was still on; I gave it full throttle to prevent it from turning over or landing too hard. The plane lifted up ten feet then settled back down. Eight men held it while we tied it down.

That night it and 80 other aircraft outside were damaged. Mine lost its left wing when it twisted downward. It took a week to be repaired. It had been tied tail to the wind. Those tied into the wind literally flew into trees and buildings.

After delivering a Taylorcraft to Fullerton Airport, I traveled to Dallas, Texas, for a Luscombe, then started back. Fuel low, I landed at a small air strip in Texas. In those days I tried to give business to the small airport operators.

Unfortunately, I'd often find that the single operator was up with a student or out of fuel, have no transportation to town and nothing to eat at the field, or simply be closed for the day with no outside telephone, forcing me to fly to the next flight service. Reluctantly, I had to abandon them and head for large municipal airports as first choice.

At this Texas air strip, I found one of the mechanics that I had worked under at Quartzsite, Arizona. He and his family lived in a small house on the edge of the field. They were friendly to me. After refueling the plane, I flew on to California.

Then I was back in Texas ferrying another plane. This one overheated and ran exceedingly rough. I was near the air strip and decided to ask the mechanic I knew to work on it. While he drank heavily and used profane language, he was a good mechanic.

He and his assistant took off the cowling to look at the engine. At my request they listed things to be done, cost, how long it would take. Then I called my employer, Les Farrar. Les later said I was the only ferry pilot he employed who had all the information necessary for a decision before calling him.

The mechanic said he had to fly a plane to San Antonio for licensing, but his assistant could start on my job. He and his wife insisted I stay at their home. I preferred to stay at a hotel, but it was convenient there in overseeing the repair work, and the living room couch looked comfortable.

The mechanic flew away and promptly got drunk in San Antonio. Two men from the town, probably the plane's owners, drove there and brought him back.

The mechanic's wife talked to the men while he staggered into the house. This fellow was well over six foot tall, a great, ugly man. He saw me in the living room and let loose with a flood of profane words. He told me what he was going to do to me when he found me alone.

I was mad and scared and went outside to his wife. "Did you hear what he said?"

"Yes, I did." She was miserable and beaten, tears welling.

"What are going to do about it?"

"Well, I don't know."

"That makes me really mad!"

I stomped into the hanger to see how fast the airplane could be put together. The assistant said it would be ready in the morning.

The drunk stumbled into his bedroom and lay on his bed. His wife told me there was a dance in town that evening and some of the young military men were coming from Brownsville. It sounded like fun, so off we went in their truck.

A young Army captain and I danced together most of the evening. Meanwhile, the drunk awoke, called a cab, and came into town. He continued to drink. Finally, his wife took him home in their truck. Both were big, hefty people; there was no room for me.

The young captain said, "Don't worry about Claire. I'll take her back in a cab."

We followed and on arrival the wife asked the captain to come in the house. He agreed; perhaps he was worried about me. The drunk was asleep in the truck and we left him there.

His wife said, "I guess he's going to sleep all night in the truck." She turned to the young man. "I'll tell you what. Claire and I can sleep in the bedroom. You take Claire's bed in the living room. It's awfully late, anyway."

The captain agreed. I went to bed, the captain went to bed, and the old gal went out to check on her husband. The drunk woke up and came into the house.

He opened a door to the bedroom from the back of the house. I did not move, not knowing what this ugly old man might do, whether he knew I was in that bed. Against outside lights he loomed twice as large as he was.

He stumbled through the bedroom to the living room. I heard the covers of the couch bed being carefully pulled back... and all hell broke loose.

I turned on the light in the living room. The captain's face was white, the drunk's face was red, the air was blue with foul language.

The captain called for a cab. Everyone settled down. As soon as the plane was ready that morning I flew away.

Chapter 4

In 1945 the war with Japan ended. Local flying on the coasts was allowed once again and flight schools reopened. I had logged 488 hours ferrying small aircraft and had acquired some business experience. It was time to get my flight instructor rating.

Fullerton Air Service needed a general all-around person familiar with aircraft and in July 1945 I began refueling the school airplanes, dispatching aircraft, doing general office work and whatever else was necessary. I gave the job my full attention and enthusiasm.

I suggested ways of doing things that made the work easier, offering to do it for them. I worked more hours each day than I was paid for. This was my "college", and my degree someday would be my own flight school.

This job was not just a job; I could've found one in another kind of business. But, I wanted my own flight school. I'd seen good and poor managers as a student and as a ferry pilot, but only in my handling every task on the other side of the counter could I really see management of a flight school as it was and as it should be.

The public comes first to the front desk with their requests and complaints. It is a great learning position for a future manager or owner of any business.

Most of my paycheck paid for flight time and instruction, and most of my "free" time was spent flying and studying flight instructor manuals. I lived in Santa Ana at my mother's and rode the bus to Fullerton each day.

On June 25, 1946, eleven months after beginning the study, I received my flight instructor certificate and began instructing at Fullerton Air Service. The first student I soloed was Ruth Lewis, a woman taxi driver.

Flight instructing is, I suppose, a rather odd way to make a living. It is the only career I ever wanted to follow, but most instructors plan to become airline pilots or company executive pilots, or fly overseas charter flights.

The necessary ratings for these require a long, expensive, practical and theoretical education. Flight instructing helps pay expenses while one is logging hours and practicing skills. Instructors are usually paid an hourly fee based on their hours flown, with benefits varying from school to school.

All of which implies there must be someone to instruct. This someone comes in all degrees of colors, creeds, sexes, stability, ability, character, intelligence, training and background.

In the 41 years since my first student to the moment I closed my school, thousands came from throughout the world. I cannot, literally, go anywhere on this planet without meeting one of my former students, even in the Sahara Desert, where one former black student is a tribal leader of nomads.

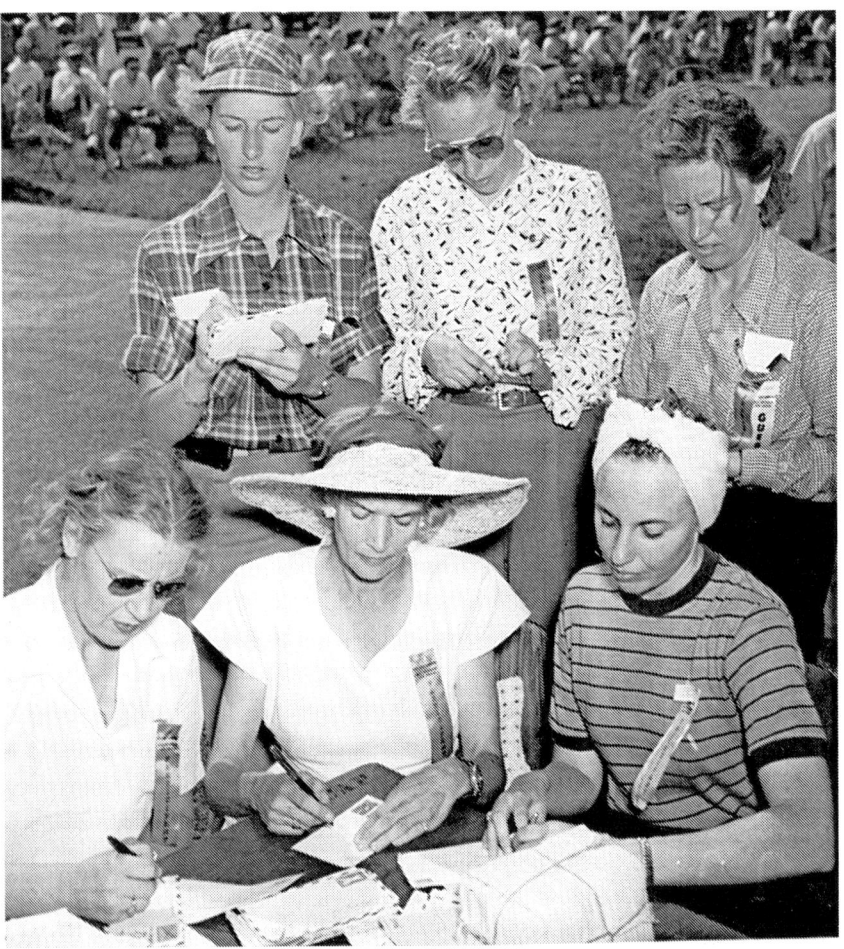

Women pilots in Miami, Florida have just flown the 2nd transcontinental air race in 1948 from Palm Springs, California. Winner was Frances Nolde (center, first row). Claire is on the right. At left is Dorothy Kravoza. Top row, left to right, Dee Thurmond, Betty Loufek, and Helen Greinke. Not shown is Jean D'Ambly. We are signing the letters we carried across the continent. Later we flew to Cuba.

Unless we keep in touch, which I do with a great many, I usually do not remember a student of a few years back...unless something happened. So if I remember you too well, and it's been a long time since I've seen or heard from you, what did you do to give me heart failure that time?

In August 1946, I began instructing for the Navy Flying Club based at Cranford Airport near Artesia. The club had two 65 horsepower Piper Cubs and claimed 200 members, but no more than 25 ever came. The two planes were well maintained and the group was wonderful to work with. I still see some of them.

Student instruction flight periods normally last for one hour except for cross-country flights. One day I took one of the students out to the practice area for two hours instead of one hour. While we were gone, two Cubs flew in formation over the airport. Shortly after that, a report came in to the flight office that a yellow and black plane (Piper Cub colors) had come apart in the air and crashed nearby.

Rumor placed my student and me in that plane: we were not back at the expected time; two planes were seen to be too close — perhaps they had collided.

As we flew into the airport pattern, we saw the crowd below. It was an ordinary day at the field; there should be no reason for it. Our plane had no radio so we couldn't ask anyone.

We landed and taxied to our tie-down space. The crowd moved towards us, led by Bill Cranford, owner of the field. He was an older man, lovable, and normally a big tease. There were tears in his eyes as he put his arms around me. People cared about me. What a terrific feeling that is.

To get to Cranford Airport, I walked five blocks to catch the early morning Pacific Electric red car on its way through Santa Ana to Los Angeles, got off about a mile from the airport and walked. The red car track was a single one and the train coming from Los Angeles each morning sat on a siding near Cypress while ours passed by.

One November morning in 1946, I was late leaving the house and ran for the train. It was one of those visibility zero, foggy mornings Southern California has frequently in the fall and winter. As I ran, an inner voice said, "Don't!" A feeling of dread seized me and I shivered but I hurried on to make my nine o'clock flight instruction appointment.

I sat in the middle of the car's main section. The awful feeling of impending disaster came again. I knew, then, that I would be hurt that morning. How? Would a car hit me as I walked along the foggy road to the airport?

Brakes squealed, the train slid. We braced for the crash. A deafening roar; metal and wood splintered; glass windows and doors shattered. Silence. Then the awful screaming began.

Tossed over seats and jammed against the front wall, I was pinned down by several people on top of me. Someone's knees hit me in the middle of my back, knocking my breath out of me. Screaming rose in crescendo. Terrible. I can hear it still.

Squirming bodies moved off me and I breathed again. What should I do? Could I help someone? I looked around. No fire, and passengers were calming. Faces were cut and bones broken, but I saw no one dead.

My legs seemed all right, but my neck and back hurt and I couldn't stand. I crawled towards the rear doors. Moving past my seat, I saw my purse on the floor with its contents scattered about. I left them there. Someone picked up my things and gave them to me before an ambulance took me away.

Reaching the rear sliding doors, I tried but could not open them. Then I felt strong arms gently lift me to a seat as the police arrived. They walked through our car saying, "Get the flares out. Another train is due." Terrified, I pictured a train racing towards us.

We learned later that the electric car coming from Los Angeles missed the siding in the fog. The two cars hit head-on, leaving part of one train under the other.

Thin wailing of ambulance sirens came closer, stopped, red lights flashed. Seventeen of us were placed on stretchers, carried to ambulances and taken to hospitals. The next day I was moved to another hospital where my mother, a nurse's aide, could help care for me.

Two vertebrae in the lower part of my back were cracked but nothing was wrong with my neck. I wore a cast for awhile, then a brace, but I could not instruct for several months. I used part of the insurance settlement to buy a 1941 DeSoto with a new engine.

After recovering, I was still able to participate in sports. But, even today, or perhaps especially today as I grow older, my back aches a bit in the morning. I returned to flight instructing at Cranford Airport in May 1947. A little while later the flying club disbanded.

While I was instructing in the Artesia area, I became re-acquainted with a man I once knew as an airport manager in the desert. Some years before, while overseas, he courted a nun. She climbed over the wall of the convent one night, married him, and came with him to the United States. When they left the California desert they came to the Artesia area.

One day he asked me if I'd like to go out to the desert to hunt. I said yes. We hopped into his little Culver Cadet, taking two guns, and flew to the desert. We shot at some wild rabbits. I managed to hit one only because the poor little thing jumped up in front of the bullet. It took some time to die, which distressed me terribly. It was the only time I have shot any creature. That is not a sport; that is no fun.

We put our guns back into the plane and took off for home. The take off was difficult. We were on an airport used by the military during the war as a practice field, but it was so hot that the little plane used all of the long runway for the take off. We barely cleared the wires at the end of the runway and headed toward El Cajon Pass and home. During the flight, the pilot decided to do a little love-making. I got upset, he got upset, I threatened to crash the airplane, he gave up.

I did not see the fellow again for many years until I was working for a flight service at Santa Monica Airport. I saw four large men get into a four-place Piper Tri-Pacer and taxi out to the runway. The little airplane was so heavily loaded with baggage and men that the tail nearly dragged on the ground. Whenever the pilot throttled back on power the tail sank to the ground.

They took off and flew through Sepulveda Pass. Near Van Nuys Airport they got into clouds where the pilot apparently became disoriented, for the plane stalled and spiraled down out of the clouds, hit the ground, and all were killed. The stall was unrecoverable; the plane was grossly overweight and far out of its center of gravity.

Sometime after all the investigations, I was subpoenaed to appear in the court case, held in Chico. Before the subpoena was handed to me the man asked if I would appear as a witness. I said, yes, but they must pay my charges as an expert witness, plus expenses.

Chico is an out-of-the-way place and not easy to reach quickly. The only way, really, was to transport me by private plane. So I rented me one of my planes, and I flew me there.

The courthouse was an old-style building. I walked into the courtroom. The jury could have sat for one of Norman Rockwell's paintings. At least eight attorneys were present, representing Piper Aircraft Company, the insurance companies, the flight service renting the aircraft, and the passengers' survivors.

The manager of the flying service renting out the plane was the man I had gone hunting with 18 years before. He seemed not to recognize me and I had no chance to speak to him.

From February to April 1948, I instructed for Airways Flying School at East Los Angeles Airport. My first student there was my brother, Deane McMillen, a commercial student. We flew Taylorcrafts; these were side-by-side two-seaters. My logbook now totaled 988 flying hours.

On April 15, 1948, my twin sister Betty set the national feminine altitude above release record for sailplanes. At the same moment, she became the first woman (and the third person) in America to fly in the mountain wave in a sailplane. She tells that story in the next chapter.

In August 1949, Betty invited me to fly with her in her two-place sailplane to set American feminine national multiplace records. None were established; she thought it was time. I had never flown in a sailplane and thought it might be fun.

The flight was to be from El Mirage, an air strip between Palmdale and Victorville in the Mohave desert, with the distance dependent upon soaring conditions.

Strapping on a parachute, I climbed into the rear seat and fastened the seat belt and shoulder harness. John, Betty's husband of two years, turned on the official, sealed barograph on the shelf behind me. He had brought the sailplane on its trailer to the marriage and my sister, the car to tow it. Betty was in front. John closed the plexiglass canopy and Betty latched it.

The 300-foot nylon towline was stretched out on the runway; the sailplane was ready. The tow plane pilot taxied to the end of the tow line. The pilot got out and snapped the line's end ring on to a hook while John placed our end of the line on the hook at the front of the sailplane, walked over to the end of one wing, and lifted it level. Sailplanes have one landing wheel, centered under the fuselage, and one wing rests on the ground before take off.

John waved to the tow pilot. We moved down the runway and in moments rose three hundred feet behind and slightly above the tow plane, out of its slipstream.

The tow pilot took us to a thermal (a column of warm, rising air) where several other sailplanes circled, all in the same direction. Betty pulled the knob on her instrument panel, releasing the towline.

The tow pilot turned and dove for the runway. Betty started circling. We circled and circled and circled, and as we rose I saw the planes at the top leave and the next ones rise. We got to the top and left it also, flying northeastward.

Betty found two more thermals and we soared on, crossed Highway 395, flew over Helendale air strip, then picked up the highway to Daggett and Baker. The air became smooth. Betty decided to land at Daggett Airport; the next airport was 60 miles farther and it served no purpose to make a rough desert landing miles from a telephone. She had no radio in the plane. Today sailplane pilots on cross-country flights carry radios and are in frequent contact with their support teams.

Betty landed the plane, rolled it close to the Daggett Airport office, and carefully lowered the left wingtip to the ground. We got out, turned off the barograph, had witnesses sign the record papers, and bought soft drinks while waiting for John to arrive in the tow car.

We set three American feminine national multiplace records that day: distance, altitude, and duration. All were broken long ago.

This has always been a special day for me. I was able not only to go along as navigator/co-pilot on a record setting flight, but it was with my twin sister. To soar like a bird was fabulous; to achieve something was icing on the cake.

In July 1950, Betty established the feminine national distance record. She established these records in the hope of creating interest among the other five American women sailplane pilots. Eventually these records were broken and others established, such as speed around a course and out-and-return flights, by women pilots such as Helen Dick and Betsy Woodward.

Many women crewed for their husbands, driving the tow cars and helping to assemble and disassemble planes, but few made flights even as passengers. Betty's husband wanted her to fly and to set records. They crewed for each other, taking turns flying the sailplane and driving the tow car.

Betty flew in the 1954 National Sailplane Contest at Lake Elsinore — the only woman entered — just three months after her second son was born. David, two years old, and Richard, three months, were taken care of by their father, who simply took them with him while driving the tow car following Betty on her cross-country flights.

Let's back up a little in our chronology. A major event occurred for women in aviation, although no one at the time could foresee its consequences. In June 1947, the Florida Chapter of The Ninety-Nines held their first air show in Tampa. That chapter invited other 99s to fly to Tampa.

Dianna Converse Cyrus (Bixby), owner of a Douglas DC-3 and a Douglas A-26, proposed that California 99s race to Tampa. She entered her A-26 and Carolyn West, her Ercoupe. It was to begin from Clover Field, Santa Monica, start of the historic Women's Air Derby to Cleveland, Ohio, August 18-26, 1929.

Fog was forecast at Santa Monica the next morning and the "race" was moved to Palm Springs. Dianna's plane developed engine trouble and never started. Carolyn and co-pilot Bea Medes were flagged off and it was not until they arrived in Tampa did they know that they had raced alone.

Since Carolyn and Bea had raced against the Ercoupe's advertised air speed, the "Amelia Earhart Memorial Race," as it was named, was declared on and they were announced the winners of the race. The first annual transcontinental air race for women had been run. And so began what was to become the finest series of air races ever held.

In 1977, 31 years later, the final All-Woman Transcontinental Air Race (Powder Puff Derby) was held. Called "Thirtieth Annual Commemorative Flight", it followed the same route as the first, Palm Springs to Tampa. In this final event, 150 aircraft carried 331 pilots and passengers.

Through all the years, only in 1974 was the race not flown, and this because of a national "fuel crisis".

Statistics published in "Powder Puff Derby, The Record," edited by Kay Brick, state that "a total of 4,400 contestants flew 638,536,800 miles, equivalent to more than 225 times around the world at the Equator without a fatality." We must add to the miles flown those needed to get to the Start by each race pilot, and the miles flown to get home.

This accomplishment came through a complete dedication to safety, which required upgrading of flight skills as planes became more complex and airports and flyways more congested.

For three days before each race, special lectures were given on wake turbulence, density altitude flying, desert survival, medical aspects of flying, engine care for long life and efficient operation, and more. Racers also flew instrument trainers under instructor supervision.

Dianna Bixby, whose idea started the Derby, never flew in it. She did, however help. She was vice-chairman of the 1952 Start Committee, and Chief Timer at the start for the 1954 race.

She became extremely busy flying co-pilot for an air freight line, setting records in her A-26, marrying Robert Bixby and having two children, Lillian and Robert. Robert is now an airline pilot.

Dianna was the first American woman to receive the airline transport rating, and was the nation's first woman airline captain. Many women briefly had held co-pilot seats on small airlines only to be dislodged by male refusals to work with them.

In 1943, Dianna married John Cyrus, an Air Force pilot. She learned to fly, became a flight instructor to add flying time, and got all the flight ratings needed to become a professional.

Her husband was killed in Europe in an air battle in 1945, but Dianna kept on flying and became co-pilot for a freight line.

She married Robert Bixby in 1948 and in 1949, they went into charter flying.

The Flying Tigers line began operation in 1949. Dianna and Bob leased their two DC-3s to it and continued as the pilots, with Dianna becoming the first woman airline captain. She flew the Newark-Boston segment of the Flying Tiger Air Freight system as a captain.

This, actually, was following the tradition of her New England family sailing and fishing background. Her grandmother, Mary Converse, was believed to be the only woman of record at that time to hold papers as a full captain in the Merchant Marine.

The story is told that Dianna sold tickets for the charter flights, which the passengers knew was a woman's job, then she helped them with the baggage and got them properly strapped into their seats, which they knew was a woman's job, but then Dianna closed the DC-3's door, latched it, and strode up to the pilot's cockpit. Passengers who had flown with her before took delight in watching the expressions on the faces of those who had not and now realized they were to be piloted by a woman!

Dianna was five foot two inches tall, weighed about 110 pounds, was blond and blue-eyed, with a charming smile that could quickly turn into an infectious grin. Surely she could not pilot such a big plane. But she did, of course, to the men's delight as well as relief.

Dianna flew the A-26, the "Huntress", on several runs for speed records in 1947. One of these was broken in 1991 by her son Robert Converse in a P-51D Mustang, "Huntress III". I can imagine her smiling in delight at her son's successes.

Dianna was preparing her A-20 for a round-the-world flight "to accept Amelia Earhart's challenge and to complete that flier's attempt", when, on a flight to Mexico on January 2, 1955, to pick up a load of fresh vegetables for the Los Angeles market, she ran into heavy weather and out of fuel. She died just three days before son Robert's second birthday. She was a wonderful person and we miss her.

In 1948, the Florida Chapter of The 99s again organized an all-woman air show, this time in Miami, and issued invitations for women to fly there. In California, a race was organized and named the Jacqueline Cochran All-Woman Trophy Race; Cochran provided $1500 for prize money.

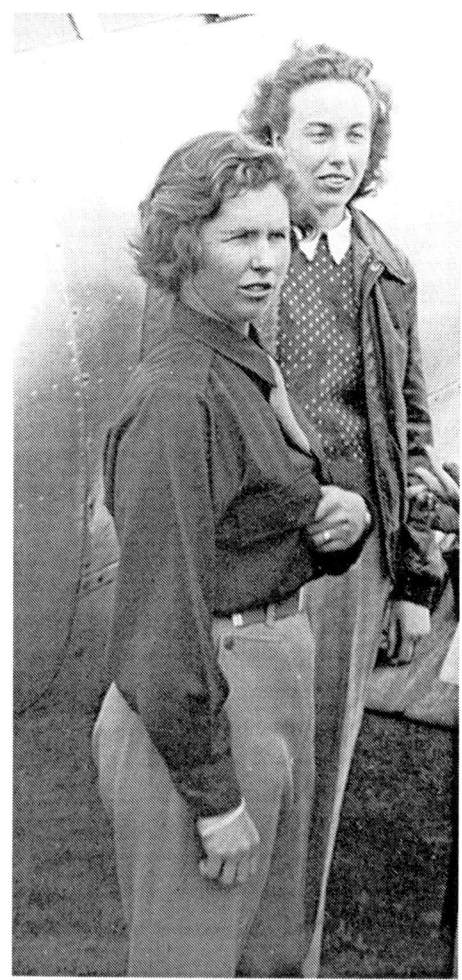

Dianna Converse Cyrus Bixby and Claire watch as pilots arrive for a 99s meeting (about 1946). At a later meeting in 1947, we learned the Florida chapter of The 99s had invited us to their airshow. Dianna said, "Let's make it a race." And so began the 30 years of the Annual All-Woman Transcontinental Air Race (Powder Puff Derby).

Photo by Betty Loufek

It was scheduled to leave from Clover Field, Santa Monica, but fog again interfered and the six planes entered were flown to Palm Springs the afternoon before the race. Planes were a Ryan Navion, a Bellanca Cruisair, a Stinson, and three Cessna 140s.

A few rules were set: daylight flying only, females only, honor system for timing, with times given to racers by control towers on landings and take offs. Handicaps were advertised cruising speeds.

In May 1948, I checked out in a Cessna 140 and flew it out to Palms Springs with my sister Betty as co-pilot, for the June 1 take off. The six planes took off five minutes apart and none of us saw the others again until we reached Miami two days later.

We flew across deserts, mountainous areas, rangelands, cities, the eastern edge of the Gulf of Mexico, and finally over the swamps of western Florida and on to Miami. All six planes finished and our plane came in fourth according to the records. We were second, though, to buzz Miami Airport.

The Florida Chapter's air show was exceptionally well done, just the right length (two hours) with a mixture of flying acts, all by women, most of whom were ex-WASPs.

The Cuba Tourist Commission invited us to Cuba, so we joined the lightplane flight to Havana. We flew with a couple in their four-place Stinson from Miami to Key West.

At Key West, we attended a United States Coast Guard lecture and demonstration on sea survival. We each bought a shark knife and borrowed "Mae West" life vests to wear on the flight.

The Stinson carried a rubber life boat loaned by the Coast Guard. The C.G. was worried about these women flying 90 miles over open sea. So was the Cuban government, which placed a Cuban Navy frigate halfway between Key West and Cuba, while a Guard rescue plane was to fly between the two points.

Today, overseas flights to Caribbean islands, to Europe and around the world are so frequent that little notice is paid to them. But in 1948, lightplane flights to Cuba or like places were rare, and especially so were those flown by women pilots.

At the lecture and demonstrations, we women wore dresses and it was a silly sight to see Mae West life vests over dresses. At least the Coast Guard did not make us fasten the leg straps. Our slacks were packed and in the plane.

Twenty-five pilots in 11 planes flew to Cuba; all arrived safely. We landed at Havana Airport, taxied to tie-down spaces, took our suitcases and walked to customs in the administration building. We looked at the observation deck and saw only a few civilian men and several armed soldiers. No women, a bad sign. Were we in danger? Dictator Batista controlled Cuba at the time, but American tourists were an important economic resource.

Grim-faced soldiers motioned with their rifles for us to go through the customs door. We entered, a little frightened. When our group was through customs, the soldiers grimly ordered us to go through a certain door. We moved fearfully through it...only to be met with a big cheer from a room full of festive men and women.

Each of us was handed a rum drink. I looked back at the open door and saw the soldiers laughing.

We had a wonderful time in Cuba. I look forward to the day when the political climate clears and we can see it again as it once was.

After the race, I ferried aircraft for various people until August 24, 1948. On that date I started instructing for National City Airport near San Diego. I was one of eight instructors for a total of 15 students. We matched for the 15-minute scenic rides. I flew eight hours that week. On the first of September I quit. No one was earning even coffee money, and since I was last hired, I should be the first to leave. Perhaps the other instructors, all men, had wives supporting them.

Student flight instruction work was slow everywhere. The Veterans Administration was slow in approving individual schools for flight training education under the G.I. Bill. Where was the next generation of airline pilots to come from when the ex-military pilot pool dried up?

My aunt, mother, and grandmother asked me to drive them to Tulsa, Oklahoma. As we traveled through El Cajon Pass in my car, a man drove up beside us and tried to force us off the road. My relatives screamed. There were few cars on the road; I speeded up in hopes of attracting the police.

Whenever the other driver caught up to us, he'd pull his car towards us and I'd pull my big, heavy car towards him. This happened several times before a policeman spotted us and stopped the man. I never found out what his problem was.

We arrived in Tulsa without further incident and mother asked me to stay. She started work at a hospital as a nurse's aide and I began instructing at Brown Airport. Same problem as in San Diego — each instructor had three students, even though the school had a Veterans Administration program.

I flew more than anyone else and averaged $125 a month. I could not make enough money to get out of Tulsa. It was a bad winter with nearly constant freezing rains which did not allow consistent training flights.

One of my women students was slow in her thinking. This woman had had lessons from other instructors and apparently misunderstood one of them. I caught her looking out of the back windows of the Aeronca Champion as we made a landing one day. I asked her why and she said she was told to do that. Apparently one of the other instructors told her to look back before she made her turn on the final, so she just kept looking back.

She was so slow in thinking I had to tell her what to do far in advance of when I wanted her to do it so she would do it when I wanted it done.

For instance, in shooting landings, while on the crosswind leg, I would tell her to throttle back to cruise and to put on the carburetor heat, so that when I really wanted it done on the downwind leg, she would be doing it. While still on the downwind leg I told her to turn so that she would be turning at the right moment for the crosswind leg. Immediately after the crosswind turn I told her to turn on final, so she could have plenty of time to think about it and still have time to do it.

While shooting landings one day, we touched a little too far down the runway and turned onto the taxi strip to taxi back to the beginning of the runway for another take off. Taking into account the usual length of time it took for any message to reach her brain and then for the commands from her to reach her motor responses, just as we started up the taxiway I told her to take off. So she did.

I joined a basketball team which ranked 15th nationally and this kept me busy in the evenings. I played guard. We attended the National AAU Tournament in St. Joseph, Missouri, and our team went to the quarter finals. Women's basketball is big in the Midwest. It is big everywhere now, but it has been big in the Midwest and the South for decades. A wonderful sport.

Basketball gave me great pleasure and the exercise I needed, and took care of an age-old problem, known today as sexual harassment. Ladies have always had the problem of how to keep one's job and still elude the boss's advances. I took care of that by having "ball practice" every night. Our team traveled at night to play in various towns. The stands were always full.

We had enough of Oklahoma's winter weather. In March 1949, my mother and I returned home to Santa Ana. My flight time was 210 hours in five months, hardly food money.

Trading cold weather for hot, I went to work at Palm Springs Airport (California) that April. The old Army Air Force office buildings were still standing then, just back of the new office building. I fixed one into a two-bedroom apartment, built chairs from nail kegs, made a table, and moved a counter from another old office building, using it as a breakfast counter which also separated living room from kitchen.

My father taught me how to do this. Every child should be taught manual skills of this sort. Think of the money they'd save, the satisfaction of creating useful objects, as well as just learning how to care for themselves.

It was necessary to create this housing because in the 1940s single women had great difficulty in finding decent places to live. Matrons did not want to rent rooms in their homes to us for fear we'd be attracted to their pot-bellied, balding husbands, who they suspected were lecherous anyway. There were few apartments and few of these were available to women.

My housing was cheap, too; it cost me $25.00 a month out of my $200.00 a month paycheck.

One of the highlights for me at Palm Springs was meeting film director Henry King and actor Gregory Peck. They were returning from Florida in King's cabin biplane after filming "Twelve O'Clock High."

I instructed at Palm Springs through the summer. It was impossibly hot. Before air conditioning and swimming pools came to the low desert, the "season" began November 15 and ended April 15. Everyone left then except for a few hardy souls and the local tribe of Indians, owner of much of the area.

Compensation for the weather comes from the scenery around Palm Springs. It is truly gorgeous and varied. The town nestles at the foot of the escarpment side of the tall San Jacinto Mountains. In the winter, one can stand warm and dry on the desert valley floor and, looking almost directly upward, watch it snow on the mountain tops. Today a tram takes you from the desert floor to the top.

Out on the desert itself, not far away, are some of the sand dunes that double for the African Sahara in films.

During June, July and August, the nights rarely cool to a comfortable level on the valley floor. From 110 to 120 degrees in the day, it might cool to 90 degrees at night.

The hot, uncomfortable, "middle of the day" starts about nine in the morning and lasts until the mountains cut off the direct glare of the sun late in the afternoon. My flying hours that summer of 1949 were from four to nine in the morning and from nine to 11 in the evening.

I was checking out a student in the Piper Cub at ten one day late in April. Already the heat was nearly unendurable. We were parched and wished we and the plane were back in the hanger.

My student had just made a "practice" landing. He let it roll on and started pushing the throttle forward for another take off. I said to the student, "Let's stop for the day."

He sighed with relief and we glanced out of the side windows to check location of the taxi strip turn off...and gasped.

A huge dust devil moved swiftly toward us from that side. A dust devil in the desert is a powerful monster. It is a thermal — an updraft of warm air — that is whirling, moving with the wind across the ground. In the big ones an "eye," a vortex, is clearly seen, and we saw this eye. Devils can cause great damage to airplanes caught untied on the ground.

Grabbing the throttle, I pushed it full forward. We were still on the ground when the terrible whirling wind struck the plane. Bumping the plane from one wheel to the other, it pushed us sideways across the runway.

Frantically working the controls, I turned the plane so that it faced into the dust devil. The Cub rose off the ground as we neared the brush at the runway edge, but its forward speed was less than the devil's and we were pushed backwards, just above the sagebrush, for a quarter of a mile.

I moved the controls in every direction with no response. Finally the Cub slipped out one side of the devil as the monster moved on majestically, leaving a shaken pair behind.

Men at the field watched our battle. They jumped in a jeep and chased after us in case we flipped. I flew the plane back into the pattern, landed and hangered the Cub for the rest of the day. That had been a wild one.

Melba Gorby Beard and I wanted to go to the 20th anniversary of The 99s Charter, held in New York City in October, 1949. Melba was a pioneer pilot, a marvelous pilot, one of The 99s charter members. We could not see the sense of each renting an airplane, so we rented one together.

This was quite a concession for Melba because she never wanted to fly with someone else in the airplane. She felt the horror of having an accident and the passenger hurt or killed. And this is a terrible thing. But she carried it to the point where she would never take anyone with her, which was too bad because Melba was an excellent pilot

Melba insisted we wear parachutes. Now the Cessna 140 is two-place and will not carry much weight. Added to our luggage and us, the two 'chutes overloaded the aircraft. But we took them and Melba felt better.

Whenever Melba handled the controls on the flight, I went to sleep. This shook her; she could not understand how anyone could do that while someone else was flying, but thought I must feel she was a safe pilot. Of course I did.

Melba held an aircraft and power rating and took care of her own airplanes. She owned a Kinner Bird which her daughter Arlene now flies. Melba died in 1987. She is missed.

Upon my return to California, I worked for James L. Most on East Los Angeles Airport. By the end of the year I had 35 solos, 15 privates, and two commercials completed. I had 160 hours ferrying aircraft, with a total flight time during that period of 378 hours. My total time was 1365 hours.

While I bragged a little about having coffee with Tennessee Ernie Ford when he came out to fly, my real satisfaction arose from the fact I was earning a fair living from flight instructing. Some students even asked for me. It was important to gain a good reputation; I depended upon "walk-ins" to earn a living in my chosen field. Friends followed me for years afterward; I gave them good instruction and they paid me well. I valued the friendship itself; we had some good, good times.

Oh, I almost forgot to tell you about something that happened the previous year, in 1948, and which has followed me for 50 years. Be careful, friends, some things never fade away.

In 1948 I flew to a Southwest Sectional 99s meeting. At the airport I ran into a fellow member who lived in the area. She asked me if I had a hotel reservation. I said no.

She said, "Oh, well, I have a room reserved at the hotel tonight, and I've decided to stay at home instead of using it. You are welcome to use the reservation." Somewhat tipsy, she slurred her words.

"Fine, that'll be just great." I took the room number, registered, and changed clothes for the dance that night.

Late in the evening I returned to my room, unlocked the door, walked in, and heard snoring coming from my bed. I sneaked into the bathroom, got my overnight kit and all my other little items, sneaked back out again and got myself another room.

I did not know who it was, did not take time to find out, and it did not matter, anyway...until 15 years later, when a group of us were rehashing old times and experiences.

In the group was Gene Moskow. Gene said that in 1948 he had been flying through San Joaquin Valley when he decided to join his friends at The 99s Southwest Sectional dance and he'd had a really good time.

Gene said that when he arrived at the airport, he did not have a hotel reservation. He saw a woman friend there. She was tipsy, and said, "Gene, I've decided to spend the night at home and won't use my reservation. Why don't you take my room?"

So 15 years later I learned who my "roommate" was. Now whenever Gene tells the story, he tells it as though we really were roommates. When I tell it, we shared the same room, but I had it the first half and Gene, the second.

Chapter 5

Claire sent me a postcard from Quartzsite, Arizona, telling me that the flight school was moving to Independence, California. I quit my "Rosie the Riveter" job at Douglas Aircraft Company subplant in Santa Ana, emptied my bank account, and took a bus to Independence, in Owens Valley between Lone Pine and Bishop. I had visited Claire once in Quartzsite and had a good time, but the flat desert was not what I like; I like mountains. The high, beautiful Sierra Nevada range is on the western side of the valley and the White Mountains, on the eastern side. I loved it.

A few days later the school owners arrived and on June 8, 1944, I started the private license course in a Piper Cub. I passed my private test on November 7, 1944.

While working on my private license, I saw an advertisement in the local newspaper asking for help at a small weather station operated by two elderly sisters.

It was one of those backyard stations where wind, temperature and relative humidity were measured and the statistics were phoned into Los Angeles. The sisters had done it themselves for many years, but now needed someone in the middle of the day so they could leave together for a few hours.

The Weather Bureau inspectors, on their next visit, told me that the station was to close as they didn't need it anymore. However, the station in Tehachapi, a small town in the mountain pass between Bakersfield and Mojave, was to be changed from Army to civilian. Would I go there as a full-time weather observer?

The Weather Bureau was in transition after World War II, opening and closing stations frequently. I returned to Owens Valley early in 1947 to work at the new weather station at Bishop Airport. I continued power flying and added sailplaning.

Flight operations on Bishop Airport were run by Harland Ross and Bob Symons. Ross was a master designer of sailplanes, and both of them were excellent pilots.

Symons owned a Pratt-Read sailplane, a side-by-side trainer built for the Army and sold for surplus after World War II. He instructed me in it on auto and aircraft tows, thermal and ridge soaring, and the art of cross-country flying.

Many sailplane pilots came to Bishop to fly the updrafts and thermals along the sides of the mountains. When the south wind blew, anyone could get Silver "C" duration time (five hours) by simply circling the bowl rim on Black Mountain. In March of 1948, I circled in it for six hours to be sure of the time.

The 1947 National Soaring Contest was in Wichita Falls, Texas. A sailplane pilot asked me if I'd help crew.

In Wichita Falls, the pilot asked me to drive the tow car. I followed the sailplane as closely as possible. When it landed we were usually right there. The two young men with me would jump out, disassemble the ship, and load it onto the trailer.

At the dance held after the closing ceremonies, I met aeronautics engineer and sailplane pilot, John Loufek. Born in Iowa, a graduate of Purdue, he was employed by an aerospace firm in St. Louis. In September he was coming to Pasadena to work on his master's degree at California Institute of Technology.

We towed John's two-place tandem Laister-Kaufmann 10A sailplane to Bishop with my old Dodge coupe and stored it in the hanger on its trailer for weekend flying. We married in November 1947 and to capsulize the next 27 years, we captured a number of sailplane trophies and records, and produced four children: David, Richard, Diane and Roberta.

Paul MacCready, Jr, was attending CalTech at the same time as John, and often came to Bishop along with other pilots.

Paul was a superb pilot, a brilliant tactician, who won the 1949 Nationals and later an international soaring contest. He told us that someday he'd build a man-powered plane. When strong, lightweight materials became available, he did just that.

John continued his studies at CalTech and I remained in Bishop at the weather station. He motorcycled up on weekends.

In February of 1948, Ross and Symons came into the weather station and asked, "Would you like to be a national champion?"

"Sure." The LK10A was available to me during the week, leaving it for John on the weekends.

I made many flights down the valley learning more about slope and thermal soaring and the art of interpreting the weather for soaring. Once I flew silently for 30 miles down the middle of Owens Valley under a line of stratocumulus clouds without making a single turn. Could this be what Heaven is like?

Symons and Ross had bought a European book on standing wave flight after seeing the enormous lenticular clouds over Owens Valley. We three studied the mountain waves, made visible when enough moisture was in the air.

The two men began making flights into the waves with power planes. They found the lift under the waves to be tremendous.

Finally it was my turn to make a wave flight — a record flight. It was April 15, 1948.

Harland Ross and I rolled my sailplane onto the Bishop Airport runway. Ross lowered its left wing to the asphalt. A 300 foot nylon tow rope with a large ring at each end lay stretched out in front of us.

The length of the lenticular cloud told us that the wave was terrific. With any luck I'd make Silver C distance and Gold C altitude. If I reached Inyokern I'd have the American women's national distance record.

The Silver C, an Federation Aeronautique Internationale award, is issued in America by the Soaring Society of America. In 1948 this award required minimum flights of five hours, 31 miles, and 3200 feet above release. I had completed the duration and altitude requirements while slope soaring on earlier flights.

Until 1938, only eight American men had received Silver Cs. In 1937, a team of German sailplane pilots, which included Hanna Reitsch, the finest woman pilot in the world, flew with the condors of the Andes to learn how they soared in thermals.

Thermal soaring was done on the east coast by German pilots who emigrated to America in 1930 and set up a soaring location at Elmira, New York. But those thermals were weak and a flight of any distance consisted more of slope soaring over the forested hills.

Richard duPont, one of only 250 sailplane pilots in America in 1934, set a world distance record of 154 miles. In contrast, at the 1969 contest in Texas, on one single day, 19 pilots made 400-mile flights; the best of the day was a flight of 527 miles.

Traveling to other countries to teach the thermal technique they had learned by watching and flying with the condors, the special German team came to America in April 1938. Delighted American pilots, quick to pick up the thermal techniques, were able to leave the slopes and soar hundreds of miles over flat country.

In the next two or three years, 32 Silver Cs were awarded. From 1945 to 1948, 49 men and one woman (Virginia "Ginny" Myers Schweizer) got them. A total of 90. I wanted the second Silver C awarded to a American woman, and to be among the first 100 Americans to receive it.

Gold C required minimum flights to 9843 feet above release, and 187 miles. By 1947, several American men had Gold Cs.

Exactly ten years after the thermal instruction by the German team, we at Bishop pioneered standing wave flights in America. On this, my first wave flight, with neither oxygen nor radio equipment aboard, I planned to stay below 17,000 feet on the flight to Inyokern Airport, 120 miles south of Bishop.

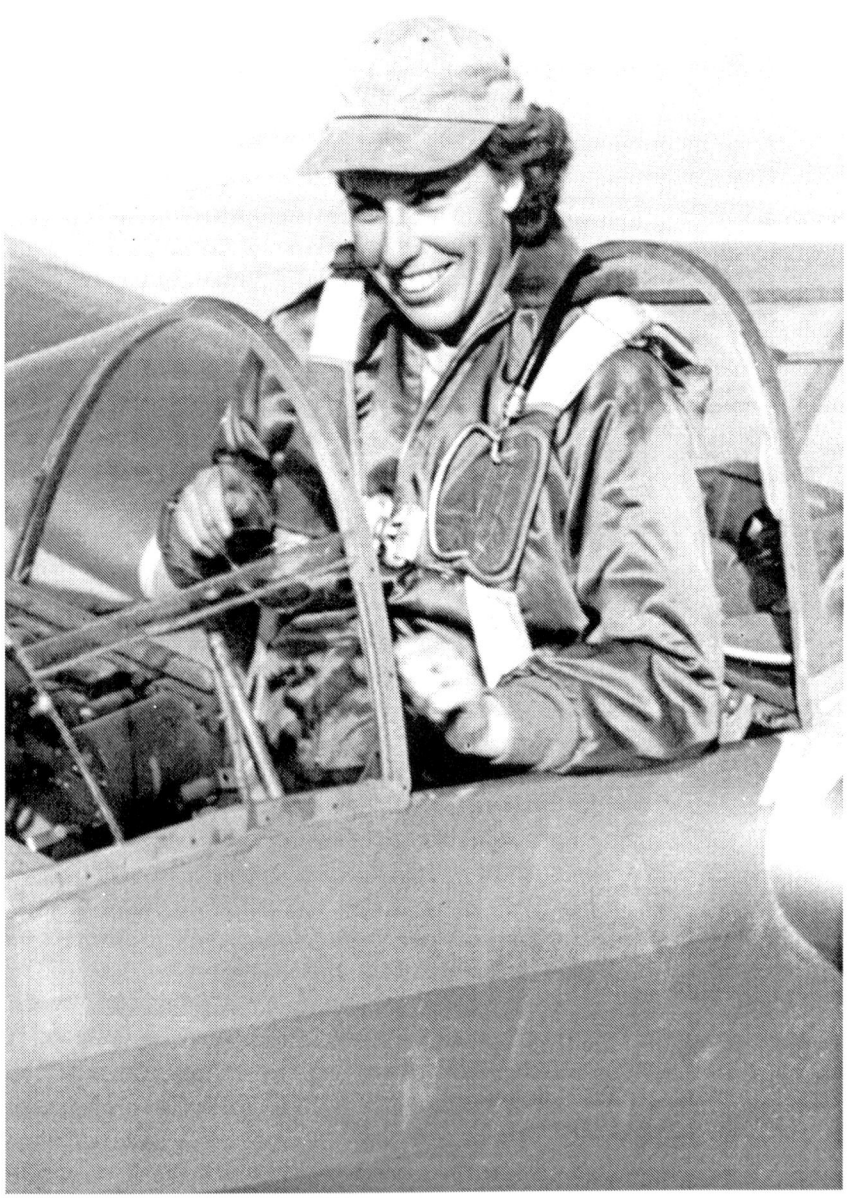

Betty Loufek is in the family sailplane, a Laister-Kaufmann two-place Army trainer. Declared surplus after World War II, this and many others like it allowed pilots to enjoy silent flight. The LK-10A was a medium-performance model; a good trainer.

Photo by John E. Loufek

At the 1947 National Soaring Contest in Texas, Ginny had soared 94 miles and rose to 7200 feet. I planned to better each by a good margin.

Bishop is at the 4000 foot elevation at the north end of Owens Valley. The Sierra Nevada range lay west. The White Mountains lay east. Both ranges are generally north-south. At Bishop, the valley is 12 miles across, becoming narrower to the south.

The valley floor is arid, covered with desert brush and rocks. The Owens River flows down it. The 15-mile-long Owens Lake between Lone Pine and Olancha, is today nearly dry.

Seated in my sailplane, I snapped on a parachute over my heavy jacket and fastened the seat belt. Ross secured the official, sealed barograph on its shelf behind me and turned it on. Another barograph was in the BT-13 tow ship piloted by Bob Symons, now taxiing slowly past. He also had my record forms.

Symons climbed out of the BT-13, now at the far end of the tow rope, snapped its ring onto the ship's hook under the tail. Ross put our ring on the plane's nose hook, walked to the end of the wing on the ground and lifted it, a signal to the tow pilot that I was ready. I shut and latched the clear plastic canopy.

The BT-13 moved smoothly down the runway, stretching out the nylon rope. A tug sent my plane following swiftly. I guided it a little above the tow ship's turbulence.

We neared the rocky slopes of the White Mountains. When the tow's orange wings waggled I pulled the towline release bar. The altimeter read 6800 feet above sea level; this was 2800 feet above the valley floor. We were three miles northeast of Bishop.

I slowed the plane from 80 to 50 miles per hour in a climbing turn away from the slope, then lost 300 feet in sink before the Robinson variometer showed more UP than DOWN as I glided near the slope. The Kollsman rate-of-climb showed 200 to 400 feet per minute UP.

I slope-soared to 11,000 feet along the Whites. A southwest wind blew steadily, creating the standing wave as well as a good slope wind on the Whites. Now I must cross through the downdraft to the valley's center to reach the wave.

The stationary lenticular clouds were visible starting at about 35,000 feet. Vertically rising air condensed its moisture at its leading edge saturation point and evaporated, raggedly, at its trailing edge as air sank. These waves continued downwind for several hundred miles.

Waves are created in the lee of any hill or mountain by wind flowing over it at about 25 miles an hour or more.

Today, I sometimes see long beautiful waves in the sky near my home in Ventura County, California. Oh, how they beckon.

I reached 12,500 feet, turned, and raced through a 1500-foot-per-minute downdraft. At 10,000 feet I entered the stratocumulus roll cloud at the point where it rotored upward. I slowed the plane and fought to keep it upright in the turbulence.

The sailplane stalled and spun one turn as it moved upward. Then the air smoothed as the plane rose into the wave itself.

I climbed at 500 feet per minute to 14,000 feet. Turning south over Owens Valley, I flew parallel to the Sierra Nevada while increasing the airspeed to 65 miles per hour. A 1500-feet-per-minute climb rate sent the altimeter through 16,000 feet. At 75 miles per hour, I was still rising at 1400 feet per minute and sailed through 17,000 feet; the Gold C altitude was mine.

As the altimeter wound through 18,000 feet, I increased the airspeed to 90 and stopped rising at 21,000 feet; at 95, the ship lost altitude slowly. Gliding down to 20,000 feet, I leveled off and sailed effortlessly at 90 miles per hour for 35 minutes. A glorious silent flight above our beautiful western land.

It was far too high without oxygen. Drowsy from hypoxia, I forgot to keep the sailplane centered in the wave.

My friend Paul MacCready, Jr., as he loaned me his parachute for the flight, had warned me about hypoxia. If my fingernails turned blue, I was in trouble and must lose altitude immediately.

My fingernails were turning blue. I pulled my glove back on. I felt great...they must be blue from the cold.

On my right far below were the Sierra Nevada peaks. On my left, the Whites, and beyond them, Death Valley. Independence was now behind me, 39 miles from Bishop, more than Silver C distance. Silver C number 91 was mine. And, I was the first American woman to fly the wave.

Lone Pine was visible ten miles ahead. The wave turned slightly, following the Sierra Nevada contour; I flew straight.

I was cold and tired; my head ached. The altimeter needle crept up; I moved the control stick gently forward.

The sailplane dropped off the wave's back edge into 3000 feet per minute DOWN. In two minutes I found myself staring UP at the Sierra peaks. With the airspeed at 80, I raced through the downdraft towards the wave as the plane sank rapidly.

The wave was above me and out of reach. Gliding swiftly to the Sierra slopes, I circled at minimum airspeed in light thermal lift to 12,000 feet. I left to reach the wave, only to sink to 9000. I turned back to the rock-strewn slopes.

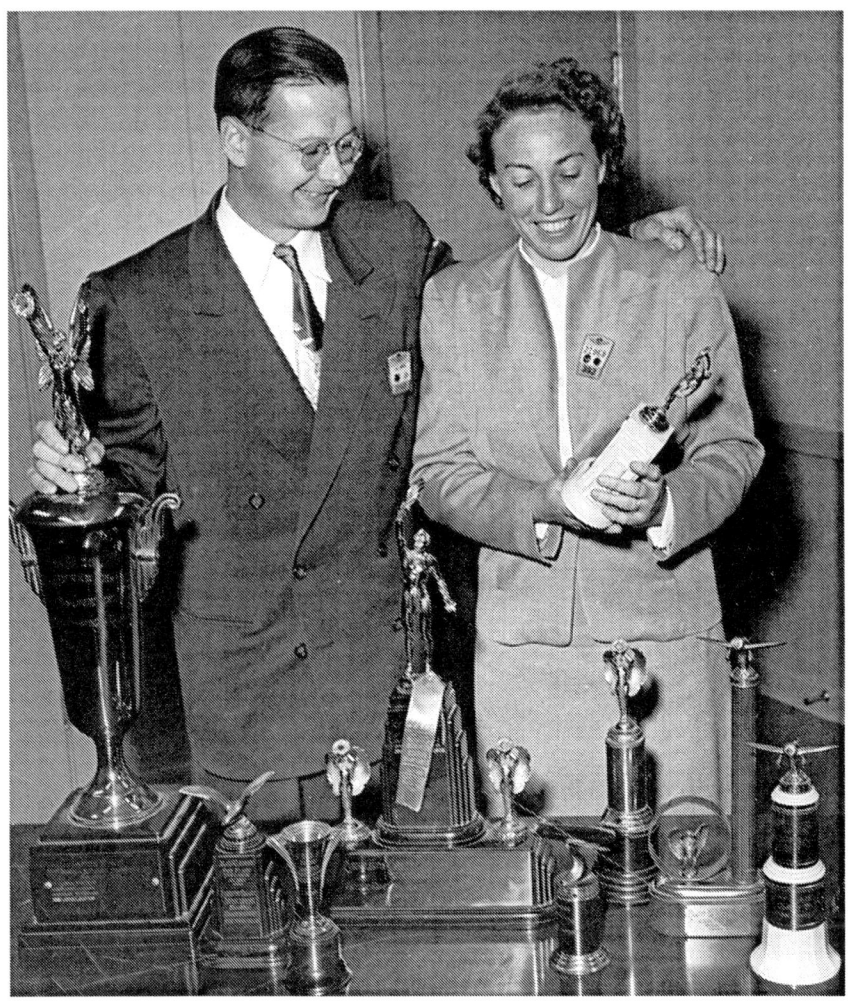

John and Betty Loufek display some of their trophies won in sailplane meets. Not displayed are four private trophies: David, Richard, Diane, and Roberta.
Photo courtesy of Douglas Aircraft Company

Lone Pine slipped behind; Owens Lake was ahead. I circled up to 11,000 feet in light thermal lift. Leaving the thermal, I picked up airspeed and raced in one last try towards the wave, only to plummet to 2500 feet above brush and rocks at the north end of the lake. I turned towards the west side of the lake and Highway 395.

Near the shoreline I gained 500 feet in light lift, then lost it while circling to find more.

Olancha Airport was at the south end of the lake. Gliding at minimum sink, I hit a downdraft that cost another 500 feet. The highway was lined with desert brush and boulders. Beyond the highway, tiny fields were surrounded by high fences, trees, telephone and power lines. Cows and horses grazed in the one large field.

Houses and trees were just 400 feet below me as I glided past the airport runway on my left. The windsock was hanging limply. I saw no air traffic. At 200 feet, I picked up airspeed, made a wide sweeping turn, leveled the wings, and lined up with the runway.

I opened the spoilers on the wings to lose the last 50 feet, landing the sailplane lightly on its one wheel. Guiding the plane to one side of the runway, I pulled the spoilers full on and braked the wheel. The left wing dropped slowly.

After opening the canopy, I unbuckled my parachute and safety belt and climbed out. I let the barograph run a bit more for a landing trace before I turned it off.

Symons arrived with the sailplane trailer. He said that while I had completed my Silver C and the Gold C altitude, instead of setting the women's national distance record — I had flown only 78 miles — I had set the women's national altitude gained record. This was the difference between my lowest point after release from the tow plane and my highest altitude.

The National Aeronautic Association confirmed it some time later. Official calibration of my barograph trace showed an altitude gain of 14,496 feet. Flight time: two hours and 40 minutes.

The Women's National Aeronautic Association awarded me their trophy, the 1948 Champion American Woman Soaring Pilot.

In 1950, I set the American women's distance record of 124 miles. That same year, with my twin sister Claire, we set three women's multiplace records.

In 1954, at the 21st National Soaring Contest at Lake Elsinore, California, I became the Women's National Soaring Champion, receiving a trophy donated by The Ninety-Nines. I could hardly miss; I was the only woman among 36 pilots.

By 1955 all my records were broken by Betsy Woodward and others. In 1955 Betsy flew a properly-equipped sailplane in the wave to 28,000 feet altitude gained after releasing from a tow at 12,000 feet.

Among those sailplaning was 99s member Helen Dick. She also flew power planes and was in the 1954 PPD. Bertha M. Ryan, a fellow writer and a 99s member, was flying her sailplane in the desert about the same time.

Today when I hear women fliers speak of dreading a possible "forced landing" (meaning an off-airport landing), I urge them to get a few hours of sailplaning. Each landing is IT, and no instructor can pour on the power for another go around. Pilots soon learn the "secrets" of planning each approach, and the fear goes away. They may even learn the joy of soaring.

This chapter is an expanded version of my article originally published in Air Trails Pictorial, May 1949, to inform all pilots on wave flying. I also spoke to several aviation groups.

Ross and Symons held a series of winter wave sailplane camps. Within five years, pilots were making wave flights to 40,000 feet and distances of hundreds of miles in various combinations of wave, thermal, storm front, and slope soaring. Pilots began wave soaring in many places in the country.

The federal government gave a wave study contract to the Southern California Soaring Society through the University of California as military pilots needed to know about it as well as general aviation.

It was many years later, from an article on Hanna Reitsch by Bertha M. Ryan in the *Woman Pilot* magazine's January/February 1996 issue, that I learned that Hanna in 1961 had "enjoyed the excitement of flying the Wave...The conditions on the lee side of the Sierra Nevada are especially strong and known worldwide."

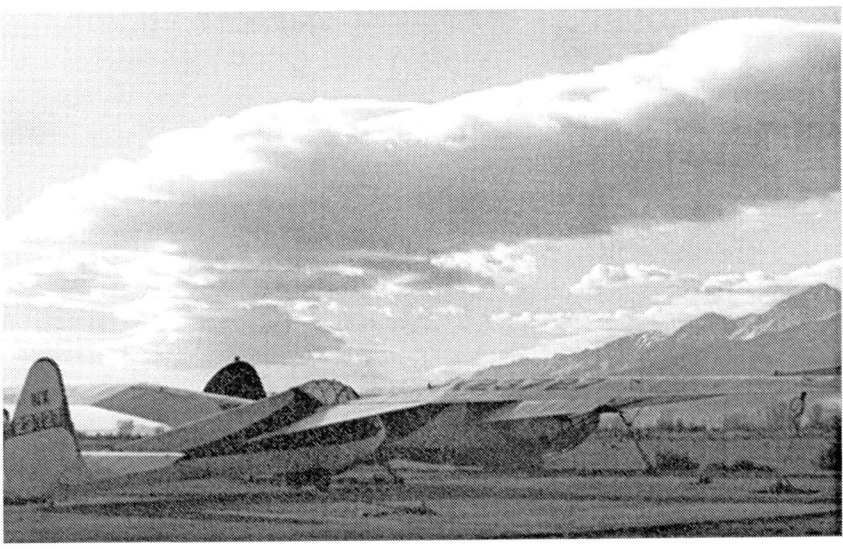

Lenticular cloud outlines the mountain wave, also called the standing wave, as it stands in place while the wave passes through it. Sailplane flights to 45,000 feet (maximum without altitude suits) have been made in like waves.

Photo by John E. Loufek

Chapter 6

Competition is addictive, I think. No matter what kind of contest it is, once someone is hooked, that person returns again and again in an attempt to better an effort or to win against a real or an imagined opponent, and to enjoy the camaraderie that exists among people of like interests. It must have started thousands of years ago; it is the reason America was first to land men on the moon in 1969. We had to beat the Soviets.

There were many short distance races and pylon contests at air meets for both men and women starting in the early 1900s. Improvements of aircraft came as a result of these contests.

The first long distance air race for women was held August 18-26, 1929, when 12 women pilots, each flying solo, left Santa Monica Airport (Clover Field) California, for Cleveland, Ohio. Comedian Will Rogers was master of ceremonies and said that "the race looked like a powder puff derby." The press gleefully called every kind of women's competition that for many years.

In 1947, as I wrote previously, the transcontinental "Amelia Earhart Memorial Race" had two planes and three pilots entered, but only the Ercoupe with Carolyn West and co-pilot Bea Medes flew from Palm Springs, California to Tampa, Florida.

In 1948, six aircraft and seven pilots raced in the "Jacqueline Cochran All-Woman Trophy Race" from Palm Springs to Miami; my sister and I were in a two-place Cessna 140. Winner Frances Nolde flew a Ryan Navion.

In 1949, 16 aircraft with 27 women pilots aboard raced in the "Jacqueline Cochran All-Woman Transcontinental Air Race" from San Diego, California to Miami, Florida. I did not fly that one but I was back for the next.

In 1950, 33 aircraft with 51 pilots aboard raced from San Diego to Greenville, South Carolina, in the first one called "All-Woman Transcontinental Air Race" by the all-woman AWTAR Board of Directors. The press called each from the first to the last, "The Powder Puff Derby."

In the 1950s a banquet speaker said when she was picked up at her hotel, "What a shame to have that other race get all the publicity

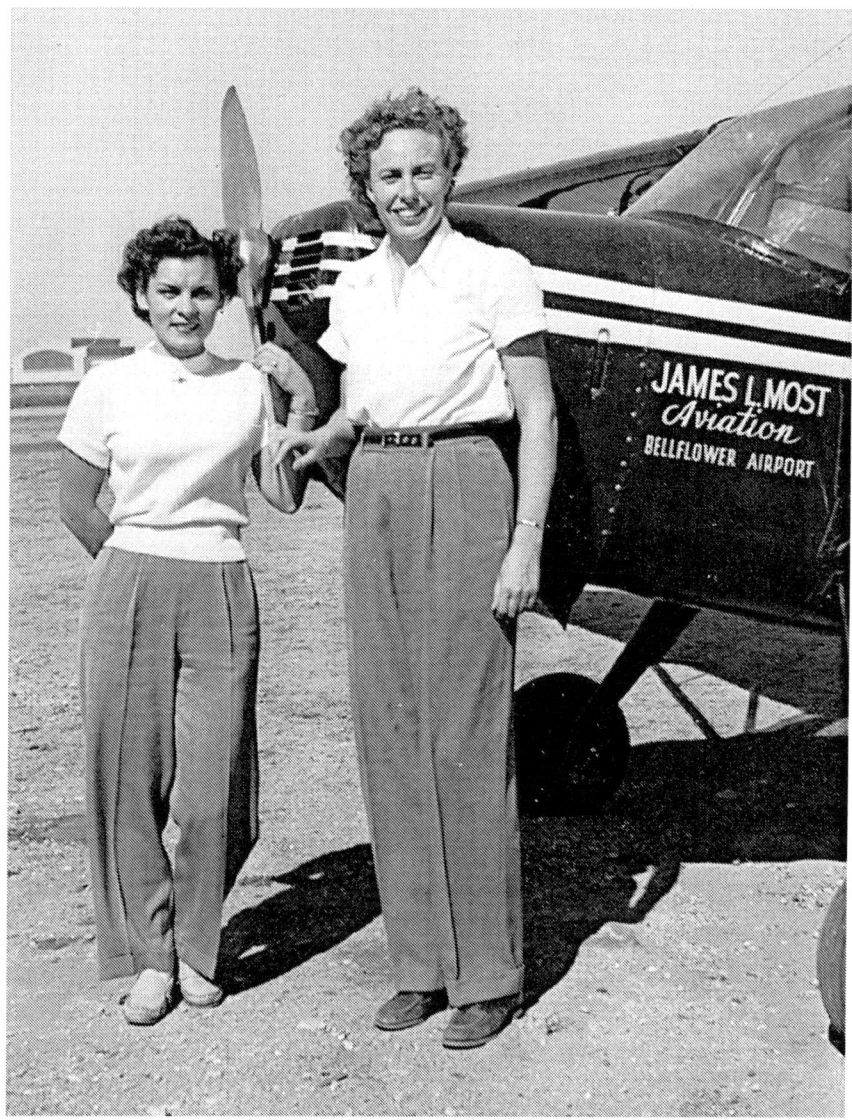

Winners of the 1951 race, Claire Walters (right) was pilot and Frances "Fran" Bera was co-pilot in the race from Santa Ana, California to Detroit, Michigan.
Photo by Betty Loufek

when you have such a great race!" The race board finally relented and let the publicist use the nickname.

Clara Davis was my co-pilot in 1950. We were sponsored by Santa Ana for the sum of $250. I appeared before the city council and asked them to sponsor us.

They were reluctant, afraid of accidents and its adverse publicity. It took quite a bit of talking to convince them, but my clinching argument was that one pilot in the race was pregnant. If she and her doctor thought it was safe for her, it was safe for us. They agreed and authorized the Junior Chamber of Commerce to give us $250 from the money the city gave them to publicize the city.

That money went a long way in those days. Our Cessna 140 was loaned to us by Art Sharp of Gardena Valley Airport, just east of Compton Airport. During the race we received free meals, free transportation, and sometimes free motel rooms. We got there and back with $40 left over.

I went to the chamber of commerce with an itemized list of expenses and gave them the $40. They about flipped.

"Boy, Claire," they said, "You're not used to being on an expense account, are you!"

Most of the pilots had sponsors who either gave money or lent aircraft, but the team that hooked a big one was that of Darline "Dottie" Sanders and Dodie Prario Cummings flying a Cessna 120. The "Chicken of the Sea" advertising department gave them matching capeskin jackets and hats shaped like tuna fish. The company arranged media interviews, and was so pleased with their performance as fifth place winners and the publicity, that the team was sent on a follow-on tour to several large cities.

Dottie Sanders went on to fly many more AWTARs, then finished second in 1971 and first in 1972, flying each of those times as co-pilot for Marian Banks.

The winner of the 1950 race was Jean Parker Rose with co-pilot "Boots" Seymour. They flew a 65 horsepower Taylorcraft, the lowest powered plane ever to win. Rose, an ex-WASP, was a marvelous pilot and flight instructor. She owned a flight school for many years. Clara Davis and I were fourth.

The 1950 San Diego to Greenville race was just great. Greenville was a medium-sized town then. The whole town knew we were there and they showed us a terrific time. There were luncheons, transportation with special drivers, all kinds of entertainment; we all had a good time.

The 1951 race was my year. Forty-four planes with 77 pilots flew from Santa Ana to Detroit. My co-pilot was Frances "Fran" Bera. This was her first race. We flew a Cessna 140 borrowed from James L. Most; we were sponsored by the Bellflower Junior Chamber of Commerce who gave us $300. We spent all of it this time.

This Flying Life

In August 1951, the Powder Puff Derby started from Orange County Airport near Santa Ana, California, and ended in Detroit, Michigan.

Fran Bera and I flew all the planes on the line at Bellflower Airport and chose the fastest. Then the Bellflower Junior Chamber of Commerce wanted to paint it dark blue and white. James Most refused then to allow us to have any of the planes on the line and said we could have the one in the hanger.

The one in the hanger was a Cessna 140 whose engine parts were scattered on the workbench. Mechanics, ram-rodded by Bud Miller, put it together and had the engine running the day before it had to be impounded at OCA (Orange County Airport). The engine had seven hours on it when Michael Walters (my fiance) and I took off for OCA in the beautiful dark blue and white "Bellflower Belles" with race number 17 on it.

The Santa Ana Junior Chamber of Commerce held an air show in conjunction with the arrival of the race teams in their aircraft. One of the show officials had called me earlier and asked me to buzz the field, and after landing, come up to the speakers' stand, since I was the hometown entry.

I told Mike that I was to buzz the runway. He said, "Well, I've never gotten to buzz any runway. I want to do the flying."

We reached OCA. He buzzed the runway and prepared to land. I said, "Mike, you're coming in too fast."

He said, "Oh, be quiet," in his usual "quiet" manner.

"Mike, slow it down, you're going to bounce." He kept up the speed to about 100 miles an hour...and bounced the plane 50 feet high right in front of the grandstand. He finally got it to stay on the ground, taxied up and parked. We were not speaking at this point.

A jeep driver picked us up at the tie-down and took us to the grandstand. We got there just in time to hear the announcer say to the crowd, "Sometimes wind and weather conditions are such that even experienced pilots like Claire McMillen have a little problem getting their planes on the ground!"

It was embarrassing and not funny then; distanced by time makes it funny now. Sort of. Years later Mike found out what it is like to be placed in that position. He had a co-pilot in his DC-7. Mike was in the left-hand seat (command seat), but the co-pilot was handling the controls. The boss was watching from the ground, and since Mike was in the left-hand seat, it appeared that he was doing the flying. The co-pilot brought it in and bounced it 50 feet high. Mike could say nothing to his boss.

Fran Bera was excited and she remembers that at Santa Ana someone told her that the only plane we had to worry about was number 17. She looked around to find out who had number 17, and then realized it was us. It was a joke.

We were flagged off and were soon over the desert. I knew the desert by heart — well, by road map, then, but Fran had done little flying over it. She was going to be a big asset as we approached the Mississippi River and flew east of it; that was her home territory. She did all the navigating when we got there.

I knew where I was all the time. "See that rock there, that's such and such." We flew strictly by pilotage: a finger on the map as we checked for landmarks.

Fran Bera had come out from Michigan just five months before the race. She was born in Michigan, learned to fly at the age of 16 in Grand Rapids, did free fall parachuting, and ferried surplus aircraft after World War II, just as I did.

She operated her own flight school in Michigan from 1946 to 1950, then moved to California.

While in Michigan she became an FAA pilot examiner for private, commercial, multi-engine and instrument, and continued this in California for 25 years, certificating over 3000 pilots. She was the first woman that the FAA designated as an examiner in Michigan. The FAA were swamped; they designated competent flight instructors such as Fran to test flight certificate applicants.

Fran Bera visited James L. Most flight school on Bellflower Airport, where I was instructing, to rent a plane and enjoy a flight around the area. She was promptly hired to instruct.

Fran is a wonderful person and we get along great. We have remained friends over these many years. We never flew the Derby together again, but in 1992 we teamed for the Palms to Pines, a race I began in 1970. It starts at Santa Monica and ends in Oregon.

We flew along in our little Cessna 140, pretty slow by today's standards, but in those days we thought it fast enough. We had to dodge thunderstorms, skirting their edges to stay as close to the course as possible.

We used our radio only to communicate with airport officials. We did not have navigational aids like we do today, and our plane did not carry the necessary equipment to use what was out there.

Fran and I worked well together on the race. I had raced before and was familiar with procedures. We had about the same amount of flight time and we both had done a lot of cross-country flying. We took turns flying and navigating.

There were 12 designated stops with time clocks and officials. They were chosen so that the planes with the least gas capacity could reach each one without having to refuel at an undesignated stop and lose time. Racers must stay overnight at one of the designated stops or be disqualified.

On the ground we had a ball. Every night we dined with the other race pilots, sharing our experiences, our frustrations when we lost time, the headwinds and tailwinds we encountered, our relief in arriving at a field before it closes because of poor weather, taking off in the calm air of dawn, the engine roaring smoothly as the earth drops away.

In that race Fran did the running to the time clock, time card in her hand. I made sure the propeller was stopped before she got out.

Betty Gillies officially became chairman of the AWTAR Board in 1952, but she had been of considerable help and influence long before that. As chairman, she made safety, always important, the dominant theme. She announced at a board meeting, "I will never attend a race pilot's funeral." And she never did. There were no racing fatalities in the Derby's 30 years.

Fran and I won the race and it was wonderful to receive the trophy at the Awards Banquet in Detroit. We had tremendous national publicity and we enjoyed it all. My fiance, Mike Walters, posed with us even though he seemed uneasy. I found out later that he had lied to his boss about why he needed a few days off. He told him his grandmother was sick. And now his picture was splashed on front pages coast-to-coast.

Coming in second were the identical twins Marion and Jan Dietrich. Some years later they were among the first 13 women who successfully completed preliminary tests in the astronaut program.

In 1960 Jan was the tenth woman in the United States to get the airline transport pilot certificate. As a corporate pilot, she flew Convairs and DC-7s. In 1968 she became the first American woman to receive an airline transport rating for a four-engine jet. Marion, also holding many ratings, turned primarily to journalism.

Edna Gardner Whyte was third in the 1951 Derby. She was a highly competitive race pilot and flight instructor whose life story is told in *Rising Above It*, written with Ann Cooper.

Edna was 89 years old when she died in 1992. During her long life she logged over 30,000 hours of flight time, won more than 125 trophies for both distance and pylon racing, and received many aviation honors. She received from Anne Morrow Lindbergh in person the Charles A. Lindbergh Lifetime Achievement Award.

At age 69 she built her own airport in Texas and, nearing 80, she built her own home near the taxiway. Edna taught hundreds of men and women to fly, continuing to do so until she lost her medical certificate at 87.

In the 1951 race and for the next several years, we could get by on little money. In the 1960s we began receiving fewer free items while race costs rose sharply. Today it is expensive to race cross-country and there are few commercial sponsors.

I have raced to cities like New York, Atlantic City, Detroit, and Wilmington (Delaware). There we stayed in one of the big hotels and people in the city and at the hotel asked, "Who are these people? What are they doing here?"

Not only did the cities and hotels not give us anything, many raised prices. In Atlantic City, for instance, the newspapers sold for 15¢ but the news vendor was forced by the hotel to sell them to us for 25¢.

The vendor told us the hotel said to raise the price to us or lose his spot, that they knew we would buy the papers because our names were in them. I do not know if they still do this; I have not been back.

I am always glad when a race begins or ends in a small town. Sometimes in the smaller towns problems can arise such as trying to find a banquet room big enough to hold hundreds of people, but those can be solved.

In large cities each person has his own problems and interests are set pretty much vertically. Rarely can he be enticed to try something new or go to an event outside of his narrow range of interests. In a small town, not only does everyone know most of the people, most join in all the town's events. And if the event publicizes the town, that is wonderful.

Some years ago a captain of an airliner mistook a runway he saw ahead for the runway he wanted but which was actually about 30 miles further on. He landed the passenger plane and then, to the captain's embarrassment, he learned the truth. He took off and brought the plane into its correct destination.

The press found out what had happened and sent the story out nationally. The small town's name was suddenly known nationally. Instead of letting the incident fade away, they had a celebration, a parade, and invited the captain. A good sport, he returned. Small towns must make everything happen locally, and visitors are welcome, especially if they help create it.

In fact, in the years of running the Derby, the officials found a problem developing as newer airplanes had greater range.

Occasionally one of the designated stops at a small town was overflown by all race pilots, none of whom needed to refuel nor wanted to risk losing precious seconds. Townspeople would be out in force, gifts piled up, timing officials on the ready. But no one stopped. Disappointment was great.

If any of you are contemplating a cross-country race of any kind, consider routing it through the smaller towns as well as have it start and or end in a small town, if possible. You might have more fun as well as a lot more cooperation.

Fran Bera enjoyed the race so much she entered 20 more times. She won it seven times — more than anyone else — placed second five times, placed third once, fourth once, and in the 21 years was out of the money only four times. She also took her turn on the AWTAR Board of Directors.

Fran believed, and I agree, that the Derby gave encouragement to women who fly. Once they knew they could fly across the continent, it built up their confidence in flying anywhere and in the use of airplanes. That was what the race purpose was, she thought. And seeing that these women can learn to fly at 40 or 50 gave others confidence in flying.

Fran was asked by one of the race pilots to ferry her plane back to California, so Mike and I flew the Cessna 140 to see his relatives in Pennsylvania and then to see mine in Oklahoma.

Mike was my student. I had taken him through private, commercial, multi-engine, and flight instructor certificates. His goal was to be an airline pilot. He reached it and retired a few years ago as a captain.

We agreed to be married in Tucson, Arizona. It had been a designated stop for the race; a warm, friendly town. My friend Maggie Shock and her husband were to be our witnesses.

We arrived in Tucson on a Sunday afternoon. People all across the country knew that we were to marry somewhere along the way. Reporters and photographers met us everywhere we stopped. Tucson was no exception. They took pictures of us at the airport. I enjoyed it; Mike did not.

The mayor had left word that if we arrived on a Sunday when city offices were closed, we were to be taken to the city clerk's home, pick her up and go to the city hall for our marriage license. Yes, my friends, being a celebrity, no matter how temporary, has its privileges. Small towns, as Tucson was, enjoy the furor. We were driven into the city where we picked up the city clerk at her home and went to the city hall.

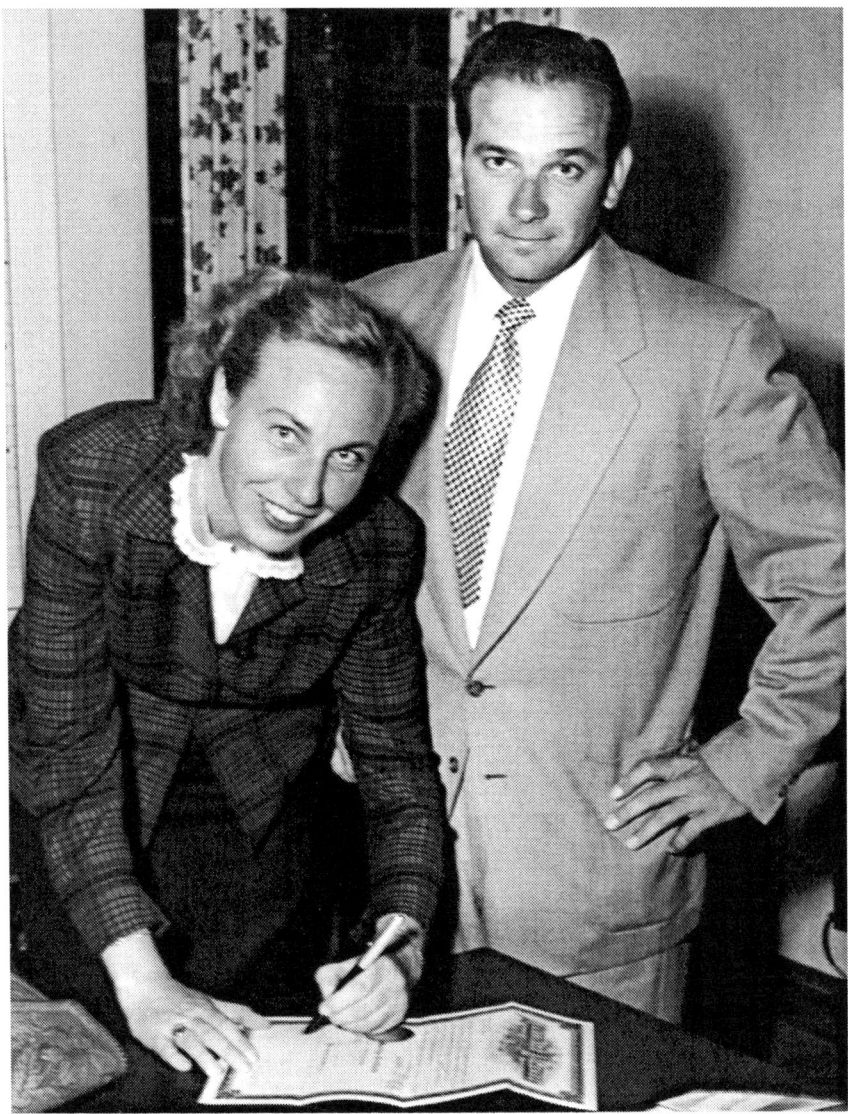

Claire McMillen and Michael Walters marry in Tucson, Arizona on the way home after Claire and Fran Bera won the 1951 air race.
Photo courtesy of Arizona Daily Star in Tucson, Arizona (1951)

A reporter and a photographer were waiting for us. They took pictures of us getting our license, and then they wanted to know when we were getting married. Mike told them it was to be the next morning. They asked him if this were true, and he said yes. I doubt he realized we had telegraphed the truth; after all, we could have, in that event, waited until the next morning, a business day, to get our license.

As soon as the pair was out of sight, Mike told me we were getting married that night so there would not be any reporters.

We went to a hotel, got our clothes cleaned, arranged for a minister and a chapel for ten o'clock. I called our witnesses.

We took a taxi to the chapel and there, waiting outside of the door, were the photographer and the reporter. Now, I do not know who tipped them off, but I suspect the press simply called the chapels that were open Sunday evening, or arranged for the chapels to call them.

Mike was furious. The press people told him to relax because they were going to take pictures and he might as well look good in them. Well, he looked terrible.

They took pictures of us signing the marriage papers in the minister's office. When it came time for us to enter the chapel, Mike told them they could not take pictures inside the chapel; they were to stay outside. We went in, our friends were waiting, and the wedding ceremony started. Out of the corner of my eye I saw the reporter and the photographer sneak into the chapel and sit down. Just as the ceremony concluded, the reporter came up the aisle, tapped Mike on the shoulder, and said, "Ah, and now, young man, what do you do for a living?" Mike came around with one big swing and started to hit him. The rest of us stopped that and after chatting with everyone for a bit, we left. And that is how I remember my wedding.

The next day I was the honor guest of the mayor and his group at a presentation of the leg prize Fran Bera and I had won flying from Santa Ana to Tucson. It was a beautiful trophy. The local radio station broadcasted the event. Afterwards I had lunch with some Ninety-Nine friends. Meanwhile, back at the hotel, Mike was sulking because I was getting this presentation and having a good time. It was his problem; he could have come.

I had begun teaching Dr. Ralph Graham in August 1951, just before the beginning of the Derby. Ten months after I married, little Mike was born.

During the months I was teaching Dr. Graham to fly, I was going to his office for examinations, of course. But, all he would talk about was flying, nothing about the baby, what to do, that sort of thing. So, when he'd come out to the airport and climbed into the airplane, I'd give him the third degree about the baby. One day, as the time neared for the baby to be born, a bunch of us went on a Sunday morning breakfast flight. Other instructors, line-boys, and students, were quite concerned. One of the line-boys asked the doctor what would happen if the baby should be born on the cross-country flight. The doctor said, "Oh, we're taking a basket, just in case!" Dr. Graham received his private certificate on June 6, 1952. He promptly grounded me. On June 30, Michael Junior was born at Long Beach Community Hospital. He had 700 hours flying time under my belt.

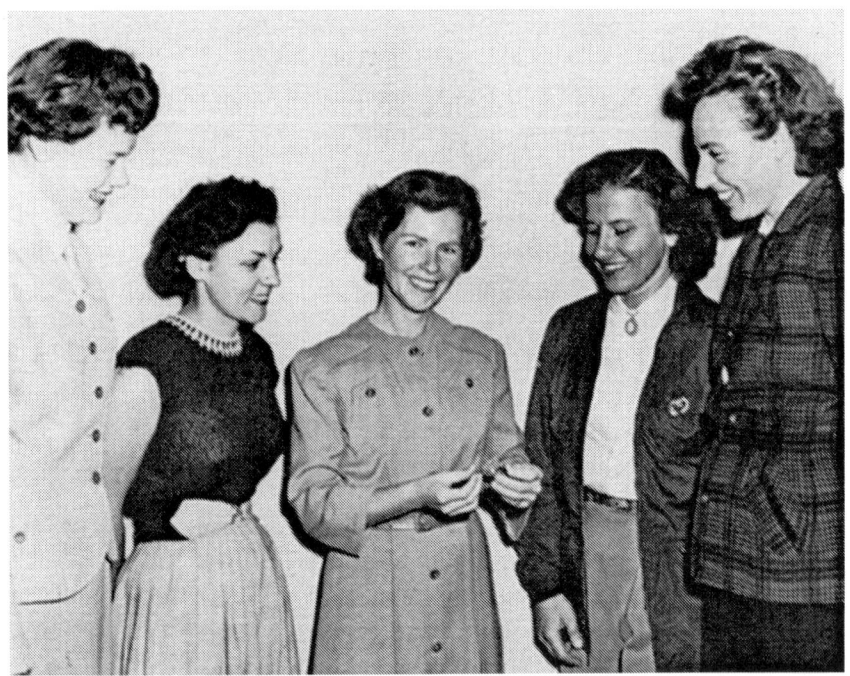

Left to Right: **Shirley Blocki Froyd, Frances Bera, Dianna Bixby, Jean Parker and Claire Walters. These four winners of the transcontinental air races (Powder Puff Derby) wish Dianna Bixby well on her planned around-the-world flight for the next year, 1955. All were also on committees for the 1954 race from Long Beach, California to Knoxville, Tennessee. Dianna Bixby was chief of the time clock crew at the starting line. Her race idea had grown like Topsy!**

Photo arranged by Betty Loufek, publicist for the race.
Photo by Wilkins of Long Beach, California

Chapter 7

After trailing around with Mike for almost three years while he started his airline pilot career, we divorced because I was pregnant again.

I returned to Southern California with our son Michael, Junior, to await the birth of my second child, my beautiful Susan, and then to start again my flight instructor career.

In May 1955, Susan was 20 days old when Fran Bera asked me to take her place at a flying school on Santa Monica Airport while she prepared for the next Powder Puff Derby. Mother had moved in with me to take care of the children.

Fran returned after winning the race and we both worked at the school. Later she and her husband moved to Long Beach, about 20 miles away, where she worked for another company.

I worked for the flight school for five years without a day off except in the summer of 1959 when I flew the Derby as co-pilot to Elly Beinhorn, a German pilot, and later when I spent three weeks in Europe as a tourist. On my return I worked at the school another full year without a day off until May 31, 1960. On June first I opened my own school.

The school owners wanted everyone to take a day off but I was the sole support of my two children, I needed to buy a home for us, and I planned to open my school. The money I earned on that seventh day was saved for those last two purposes. On my "day off" I usually flew outside students in their own airplanes.

The tailwheel Piper Cubs had been phased out by most schools some years before. This school flew Cessna 140s. These were also tailwheel planes, but side-by-side two-place, with half-wheel controls in place of control sticks. Engines were started by controls inside the cabins.

These planes had several more instruments on the panel than were in the Cubs, including gas gauges. Gas tanks were now in the wings. All planes had radios. These planes were much more comfortable to fly.

The Cessna 140 was a poor choice as a student trainer, but at the time there was not much else on the market. It was not until early 1958 that the Cessna 150 was available.

The Cessna 140 has a tendency to groundloop (swerve to one side and go into a tight skidding turn). One must work to keep it straight on takeoffs and landings. It also lands faster and rolls faster than an Aeronca Champion or a Piper Cub. When one of those slower planes groundloops, usually the damage is minor; when a Cessna 140 groundloops, it is disastrous. It flips over and someone is hurt.

The flying school lost many Cessna 140s during the years it was in business. Some of this was due to poor plane design, but the problem was intensified through poor business and flying practices. I was certainly learning how not to run a school.

The owners believed that every student was to be soloed by eight hours of dual time. If he had not been soloed by 12 hours, the instructor lost the student since it must be the instructor's fault. As a result, some instructors soloed students too soon. Disaster followed. Students groundlooped or dragged the wingtips or bent landing gears.

Each person's ability and background differ and this must be taken into account when instructing that person. Also their sex. Women are generally much more cautious than men and will agree to solo only when they are sure they are ready. Usually men want to get rid of the instructor as early as possible.

The owners pushed the instructors, giving them such a bad time that the instructors were often tense and shaken, forced to do things they would not ordinarily do. Earning a living is difficult enough for a flight instructor; under these conditions it was almost impossible. All to impress potential customers with a promise of quick solos and short courses.

One of the owners was an extremely bad pilot who drove a plane as one does heavy machinery. His partner was a good pilot but had a terrible temper, was poor at public relations, made people angry. Customers walked away day after day. It was ridiculous to advertise for customers, then insult them.

I kept some of the students from leaving by talking to them, listening to their complaints. I told them how to finish their courses and get their licenses in spite of the circumstances.

There were many accidents after I left the school. One instructor and his student were killed right in front of the office when a Cherokee 140 stalled and spun in on take off. Another instructor and his student were killed in the mountains while looking for condors.

Over the years all of their Cessna 140s had been involved in accidents, either at the home base or at other airports. All of their original Cessna 140s were wiped out, as were most of the Cherokee 140s, the planes they had bought after I left their school. Their flying record on file in Washington was the worst of any flying school in the country. They continued their destructive practices until they went out of business due to impending lawsuits some years after I opened my school.

The owners operated the business without wanting to spend any money for tires, batteries, other repair parts. A major reason for this attitude was that they had contracted maintenance instead of having their own workshop and adequate tools. In the first two weeks I worked for them my plane had two blowouts and three flat tires. Before I rolled the plane to fly, I checked everything, of course, and spoke to one of the owners about the tires. He said, "Go out and fly them if you want to work here."

I pointed out that it was more expensive to fix them out on the runway than in the shop, considering time spent and having to bring out equipment.

"Go out and fly them if you want to work here."

I'd taxi out to fly and have a blowout on the runway. He'd have to go out and fix it. Stupid fellow.

One cause of the owners' money difficulties, or cash flow problems, to use accountant-speak, was their refusal to accept money in advance for lessons. New students offered to pay in advance for flying courses. This was refused; they were to pay by the hour as they took instruction.

We instructors asked about this. They replied, "We might have to give it back."

"Yes, You might have to give some of it back sometime, but in the meantime you'd be that much further ahead."

It had been my experience that when students have money in a school account, they are more likely to have a regular schedule of flight. Students progress evenly, and instructors can plan better. And, the school has money to properly maintain aircraft.

Since I was there seven days a week I gradually found myself running the school, building it. At least, it felt that way.

When I left the school they had moved from Cessna 140s to two Champions, two Cessna 172s for instrument flying, and seven Cessna 150s, and from about 400 flying hours a month to close to a thousand. I closed the office at night, tied all planes down, made sure each was secured.

The owners left before the day was over. They called me at night at my home to see if I had tied the planes down. This irritated me; they could have stayed and helped. I told them I was not going to answer the question; to go check the planes themselves. After awhile they stopped bothering me.

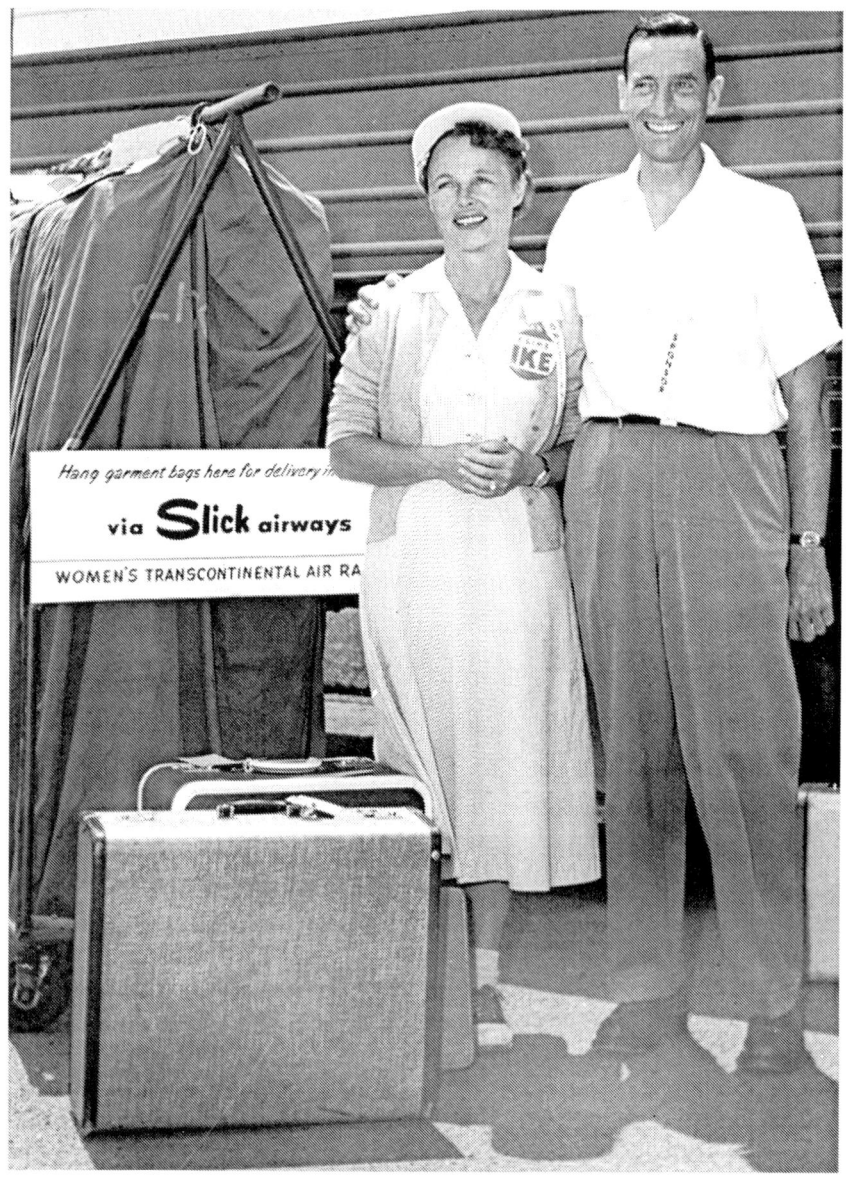

B. Allison "Bud" Gillies and Betty Huyler Gillies. Bud was on the board of directors of Slick Airways. He arranged for the racers' luggage to be flown to the terminus for each race. Betty was the air race chairman from 1952 to 1961.

Photo courtesy of Slick Airways.

I never left the field without tying down the planes and checking the office. If the planes were wrecked by winds during the night I could not fly them; if the office burned down I would have no place to work.

Finding a good instruction plane was important. In early 1958 the Cessna 150 came on the market. It was a nose-wheel aircraft, the kind we instructors wanted. Only about one in eight students do well in a tailwheel plane; almost anyone can learn to fly a plane with tricycle gear; that is, one with a nosewheel and without tailwheel or skid.

Once again I talked to the owners about buying tricycle landing gear aircraft. Now they said, "All right, if you can raise the down payment for an airplane, we'll go ahead."

Fellow instructor David Irvine and I asked a number of students if they would like to put some money down on account. In a few days we gathered down payments for five Cessna 150s. We took turns going back to the Cessna Aircraft Company in Wichita, Kansas, for them.

These airplanes were easy to fly, nice stable aircraft, and there was far less trouble with them than with any of the others.

The major problem was the attitude of the school owners. They simply caused so much trouble for the instructors that poor judgment was used and proper training was not given to the students. An instructor being bawled out in front of his student cannot maintain control and dignity. Often we could not find out what the problem was that set off the fireworks.

Having good equipment is important, but customers will even overlook rather poor equipment if the attitude of those running the flight school is good. But, customers do like newer, well-outfitted equipment and will change schools to get it.

One time the owner ordered me to sign off a student's cross-country dual flight. I told him no. I had taken that student on several cross-countries and he simply could not find his way. He was unable to use logic to get us from one point to another. I was ordered to sign him off; I refused. We had quite a row.

Another time the owner wanted me to sign off a fellow for his private license. This man could not even comply with the minimum requirements for dual under the FAA. I refused.

In 1969 I was in the Monterey Peninsula area with some other women pilots; one was a Salinas tower operator. She told me that the school still had the worst reputation and my school had the best reputation for student cross-countries.

When a student from that school filed his flight plan for Salinas, Salinas immediately notified Paso Robles and Santa Barbara to get out their plotting boards because "here they come again," and almost always the plotting boards had to be used to find the lost student and help him back to an airport.

The tower operator told me they never had that problem with my students. It was good to hear.

That particular flight school is long out of business, but there is always another one standing ready to part a student from his money. It may also end in tragedy. At a minimum, the new student will be so disappointed he may give up and never get to know the joy of flying. One must research the flying schools and talk with students and rated pilots.

One of my students was Peter Strudwick; he was born without hands or feet. It was February 1956 and he was my 96th solo student. The doctor he went to for his student medical refused him to give him a medical examination.

I spoke to the FAA and had them tell the doctor to give Pete a medical. I did not believe that doctor had any right to determine whether a person should or could learn to fly. His job was to check heart and eyes. My job was to teach him and, if he learned well, the FAA would give him his private license.

Pete passed the physical, learned to fly both the Cessna 140 and the Aeronca Champion; he was my 48th private pilot.

Pete Strudwick went on to get his Ph.D and become an international speaker and trainer in motivation, discipline, education, training, sports and health. He wrote a book, *Come Run With Me*. I saw him in 1991 in Santa Monica when he invited me to a large gathering where he was the speaker. He told the audience what I had done for him and what confidence flying had given him. He told me afterwards that he was writing another book and that that incident was in it.

A handsome young man signed up for lessons with me. I knew him as Jack Golenor. I worked literally from dawn to dusk everyday and then went home to my family; I didn't have time to go to the movies. I eventually was told his movie name was John Gavin, movie actor and leading man opposite Lana Turner and other beautiful women. His full name is John Anthony Golenor Gavin.

I did not know John Gavin as a student from any other young man, except he was extremely nice and a gentleman, and he always hugged and kissed me before and after each flight.

By August 1957 I had logged 8,100 hours, soloing 161 pilots, and successfully took students through to 82 private licenses, 26 commercial ratings, three multi-engine ratings and five instructor ratings.

Putting people under more pressure than they can bear is disastrous. When a person is unstable, pressure can be perceived even if it is not real, and the results are the same.

The landlord chose one day to designate a gas man as night manager of the flight line. He gave him no change of duties nor a salary raise, just the title. The man began to change from a quiet individual to an anxious, irritable one; began talking to himself about aliens; he didn't shave or bathe.

Several times I pointed out to anyone present, that ol' Mac appeared to be going insane. No one would listen — they told me I was working too hard.

While I was checking the tie-downs one evening prior to quitting for the day, this man angrily ordered me off the flight line. I refused to go until I had finished my duties.

The next day he came into my office, leaned over my desk, and said that he was planning to kill me; that he could get away with it by saying it was in self-defense.

I reported it to the landlord, who refused to listen. A nightmare! No one would believe that Mac had gone insane.

One night he picked up a crowbar and advanced on me. His girl friend (they were both in their late 50s or early 60s) was there and stopped him.

The next day, the woman's son came looking for her, saying she had disappeared. The deputy sheriffs caught up with the insane man at his home, and he was arrested when he took a piece of pipe to them.

Later we learned that he had apparently killed his sister and her three-year-old daughter some years before and had been in a mental institution from which he had escaped.

In December 1956, I bought a three-bedroom home just a few blocks from Santa Monica Airport, fixed it up, and turned the garage into a playroom that included a bar with a radio and record turntable.

I gave a barbecue party in my backyard, inviting about 50 people: tower personnel, flight instructors I worked with, all my friends and my friends' friends from the field. John Gavin brought a huge steak and cooked it for the two of us.

I put up tables in the backyard, brought in several portable barbecues, bought the beer and potato chips and made a salad. Each guest was to bring his own meat and cook it.

A fellow instructor had been discharged recently from the Air Force. He had been in the Philippines as a member of a secret intelligence group checking on military officers, including generals, arresting them when they were caught smuggling or dealing with the black market. A tremendous operation was going on trading "greenbacks" (American paper money) into China in exchange for gold, dope, and other items that bring high prices in the United States.

His family was with him in the Philippines while he was flying between Manila, Saigon, Macao, Hong Kong and other exotic places which are among mankind's filthiest sewers. He and his group were coming close to netting some big fish. The enemy discovered his identity and marked him for death. One afternoon he was tipped that he was to be killed that day. He went to his superior who immediately called for a C-54 transport plane to prepare for flight, ordered a military car into Manila to get the wife and children, loaded them all on the plane and sent them to the United States. Riding with him on the plane was a nightmare, fear that killers were following him. Soon after, he was discharged, severed all ties with the military, and became a flight instructor. He was coming to the barbecue.

The night before the barbecue, an old friend who came to see me whenever he was in town — every six months or a year — dropped around and invited me to go out. I said, no, that I had too much to do on the party for the next night. Maybe he would like to come to the party? He said he would.

He had just returned from the Philippines. Now he said, "Got a job for you in the Philippines, working for my boss."

"What kind of a job?"

"Flying cargo from one place to another. Manila, Saigon, Macao, Hong Kong, wherever it's supposed to go. Pays well," and he mentioned a figure.

I gasped and dropped a dish, shattering it.

"What kind of cargo? What are you carrying?"

"I don't know. I never look. Others load it while I check the weight and see it's all secure. I take it where I'm told."

"Thanks but no thanks. Nobody pays that kind of money to fly ordinary cargo. It must be illegal."

"Well, if it is, I don't know anything about it. I work for a guy named White. I just do what he says."

He showed up at the party the next night and he and the new flight instructor slowly gravitated towards each other as they circled with the other guests. As men do, they began discussing what they did for a living, and what they had done.

I glanced over at them and was shocked. They were staring at each other in horror. The flight instructor stumbled towards me, grabbed me and took me into the kitchen.

"He's come to kill me!"

I looked out the kitchen window. My friend stood there looking puzzled, hurt, shaken. Hardly the look of an assassin. "He wouldn't hurt a fly."

"You don't know! This is terrible. He's going to kill me!"

"Well, I know he's not going to kill you. Especially not in my backyard, for heaven's sake! What's this all about?" Then he told me what I have already related, about fleeing the Philippines with a price on his head.

"Your friend works for the head man, White. He's been sent to kill me."

"Not in my backyard! I'll go talk to him about this."

"What's the matter with that guy?" asked my friend. "Is he nuts or something?"

"He thinks you were sent by White to kill him."

"Oh for — I told him I work for White. I fly what he tells me to fly anywhere he wants me to, but I have never hurt anybody."

The flight instructor was not to be convinced, and sat for the rest of the evening where his back was protected and where he could watch my friend. Nothing happened to him, then or later.

My friend flew back to the Southeast Pacific to his flying job. A few months later a sniper killed him as he taxied a plane on a remote jungle field.

It was March 18, 1958. Student Jerry Bruno and I had taken off from Santa Monica Airport in a Cessna 140 and as our Palos Verde practice area seemed cloudy, we headed for the other practice area over Malibu. We were at 5000 feet and over the ocean an hour later with the lesson almost over when I told Jerry to sit back and relax a few minutes. He started to do so, then glanced upwards out of the window and shouted "Plane on fire!"

I saw it a second later. Two thousand feet above us was a ball of fire, breaking up as it spiraled down.

Switching on the radio, I called Santa Monica Tower as the fiery Douglas F4D plunged toward the sea. We were high enough to give complete information about it before the plane hit.

The Tower operator told me to orbit over the crash, that help was on the way, and to switch to 121.5, the emergency frequency. The two Tower operators on duty, William J. Crunk and Don D. Hartwick, were in complete charge of the situation. With a microphone in one hand and a telephone in the other, each started a series of calls.

They first called Douglas Aircraft Company Flight Operations at Santa Monica. Douglas Flight Operations ordered their Navion and helicopter out and called the pilot of the only other Douglas plane in the air, Roger Conant in a F4D Skyray.

The Tower operators, in the meantime, reached Los Angeles Radio Communications, who alerted the Coast Guard. The Tower called Los Angeles Air Traffic Center and it, in turn, alerted the lifeguards at Santa Monica. Usually the Center takes over the coordinating, but this time they had the Santa Monica Tower continue.

The Douglas-owned Navion and the helicopter were in the air three minutes after my call.

Conant, chief test pilot for Douglas El Segundo, said he was at 40,000 feet over Palmdale (50 miles distant) with only a few minutes of fuel, but he would try to get there.

Conant arrived so quickly at Malibu that he was able to see the parachute carrying the pilot hit the water. He flathatted across the water, dropped his tiptanks as markers and left for the Los Angeles International Airport. Fuel gone, the engine flamed out on the roll-out.

The helicopter and the Navion arrived immediately afterward. The pilots asked where I was (above them) and I answered them. Although we could talk directly, my radio seemed to be cutting in and out, so the Santa Monica Tower relayed most of our conversation.

Conant called the Santa Monica Tower, stating where he had dropped his tiptanks and where the pilot was. He was leaving nothing to chance.

The helicopter crew spotted the pilot, Leo J. Colapietro, in the water and heli-diver Leonard W. Danison dropped off from the height of ten feet and stayed with him until the lifeguard boat arrived. The pilot was pulled out and the boat headed for the dock, where an ambulance waited. Colapietro had spent about 40 minutes in the water.

When I spoke to the Tower men later, they told me that if Leo Colapietro had not been found so quickly, they were ready to assign search patterns and altitudes in order to cover the greatest amount of area in the shortest possible time.

Jerry and I returned to the field. One day during a week of filling out reports on my small part and account of the affair, test pilots from the Douglas El Segundo Division arrived at my place of work with gifts and a letter of thanks signed by all the test pilot group. One of the presents was a pearlshell inlaid compact that chimed

"The Third Man Theme." I will treasure it always. Douglas Aircraft Company president Donald Douglas Jr., sent a letter of thanks.

The Douglas pilots told us of their sincere appreciation for all our efforts. They are sure that if action had not been so swift and cooperation so complete, their friend and fellow pilot would have died.

"Pete" Colapietro had been seriously injured but eventually recovered. According to news stories, he was at 15,000 feet, two minutes out of Los Angeles International Airport, on a routine production check flight before delivering it to the Navy when the ship went out of control. The canopy tore loose and he thought it struck the tail.

Colapietro "grabbed the curtain", triggering the ejection seat. His own reactions saved him in the beginning, and his friends' and the Tower's teamwork saved him at the end.

During the last part of 1959 and early part of 1960, a Los Angeles area radio station came to the flight school owners with a deal: if they would supply the airplane and pilot to make reports on the freeway traffic and weather in the mornings and evenings, the station would advertise the school over the radio.

It was a poor deal for the flying school, but the owners agreed. They asked me if I would do it that first morning, and I said, "Sure."

A radio was placed in the plane over which the studio was to contact me to tell me when to start transmitting the report. The studio recorded it so they could play it whenever they wanted to. It was my job to take off at seven o'clock, fly over the freeways, report any accidents or undue traffic conditions.

I did it the first morning or so, then came time for payday and the money was not on the check. I said, "I'm not doing this for nothing. I have to spend two hours in the morning and two hours in the evening to do this. Either I get regular instructor's wages or I'm not going to do it."

The owner said, "All right."

The next morning I found the airplane gone. The owners were ornery people; really mean. Instead of telling me the day before that they were going to do it, they rushed out early in the morning before I got there. Then they could do it and not have to pay me. I had lost instruction time I might have scheduled, which was their loss, too.

They soon found they did not like to do it; it became a drag. They gave it back to me and paid me for it.

I took off each morning at seven, climbed to the proper altitude and made my report. Soon the station found that my transmissions were good enough to go straight out over the air. They still taped them to replay later in the day, but I gave live broadcasts for many months.

It was a foolish deal for the school. They were not paid for it. The studio did not treat the program properly. They occasionally resold the spots to many car dealers. While they mentioned the flight school and my name, they mentioned all these dealers, too; a rotten thing to do.

The bad part of it was when I had an accident report or another kind of bad traffic condition report to make, I'd call the studio on the special radio and tell them that I had something for them and I wanted it on the air because it was important to the traffic. They would say, "Wait a minute, Claire," and they would play another tune. I would say, "Come on, now, let's go." They would say, "Just a minute," then give an auto dealer's commercial. Finally they said, "Now we go to Claire Walters, high over Los Angeles, for her report."

Fifteen to 30 minutes might pass before I could give my report. It was not the public service that they had sold to the flight school and the city. They were inconsistent in the times they called me. It got so I had to listen to the commercial program because they would not give me a cue over the special radio; they gave me the cue over the radio program.

That forced me to listen to some awful music, and just when I was going to turn the program off and force them to give me a cue over the special radio, they would say, "And now we will go to Claire Walters, in the sky over Los Angeles for her report."

I tried to be in a certain spot at a certain time. Finally I called the radio people and said, "Look, I want to know exactly when the first call in the morning is going to be; you are much too inconsistent."

"The first call will be at 7:20."

"Fine. I'll position myself along the freeway at one of the interchanges and give you the true story."

One morning I was taking off at seven, and I had just reached the position above the runway where my radio transmitter and receiver was in line with the transmitting commercial station, when I heard the announcer say, "Now we will take you to Claire Walters, who is high in the air over Los Angeles, for the latest traffic report."

I was still not at the end of the runway, but I could see the San Diego Freeway so I gave them a freeway report.

The weather was bad on some mornings. It was impossible to fly over Los Angeles, even illegally; the ceilings were too low. I'd call the radio station and tell them that I could not do it. They'd reply, "Well, that's all right, we don't care, give us the report anyway!"

I'd fly over the coastline, and maybe tell them about the traffic conditions on the interchange in downtown Los Angeles while I was flying past Malibu, or I might be over the Marina del Rey, where I could see part of the San Diego Freeway.

It was not a true traffic report and I hated it. I'd listen to the signal alerts on the regular news reports and I knew there was nothing going on at the time; no traffic problems.

I complained bitterly to the radio station, but they did not care. It was a gimmick to them. They recorded what I said, then when I was in the Santa Monica office later that day the radio would blare, "And now we take you to Claire Walters who is high in the air over Los Angeles." This was maybe four in the afternoon, and it was the report I gave them at 11:00. It was dishonest and not a public service, as the station maintained.

The station reached the point where they wanted me to listen to the reports of Captain Max Schumaker and then report mine based on his. He used a helicopter to fly the freeways.

Even if I wanted to copy them I could not. There was no way to listen to his reports on another radio station and also listen to "my" station and catch my cues. But what I found hilarious one day was finding Max in a phone booth at SMO (Santa Monica Airport) trying to find out what the traffic conditions were so he could report something. The weather was so bad that even he could not fly in it.

Some years later I flew with Max in his helicopter, not long before he was killed, and we flew along the freeways. We had quite a long chat. He was an interesting person; a fine pilot. It was a terrible shame that he was killed in the collision with the police helicopter.

Another time the radio station sent me down to Long Beach to report on the tidal wave that was supposed to be coming in. Tidal waves are rarely discernible from the air; they do not get to be these giant waves like you see in the movies until they get close to shore. I flew down there. I could see the damage that had been done in one channel the night before, and I could see whirlpools out in the ocean. But I could not see any tidal wave coming in. The radio station people kept trying to get me to report to the radio audience that I could see a tidal wave coming. I refused.

Not long after this I told the owners that it was costing too much money for what they and the public were getting out of it. They agreed.

I believe I was the first freeway gal in Los Angeles. It must be even more frightening now, with all the planes and helicopters from TV and radio stations, Los Angeles Police Department, County Sheriff's deputies, and California Highway Patrol up there.

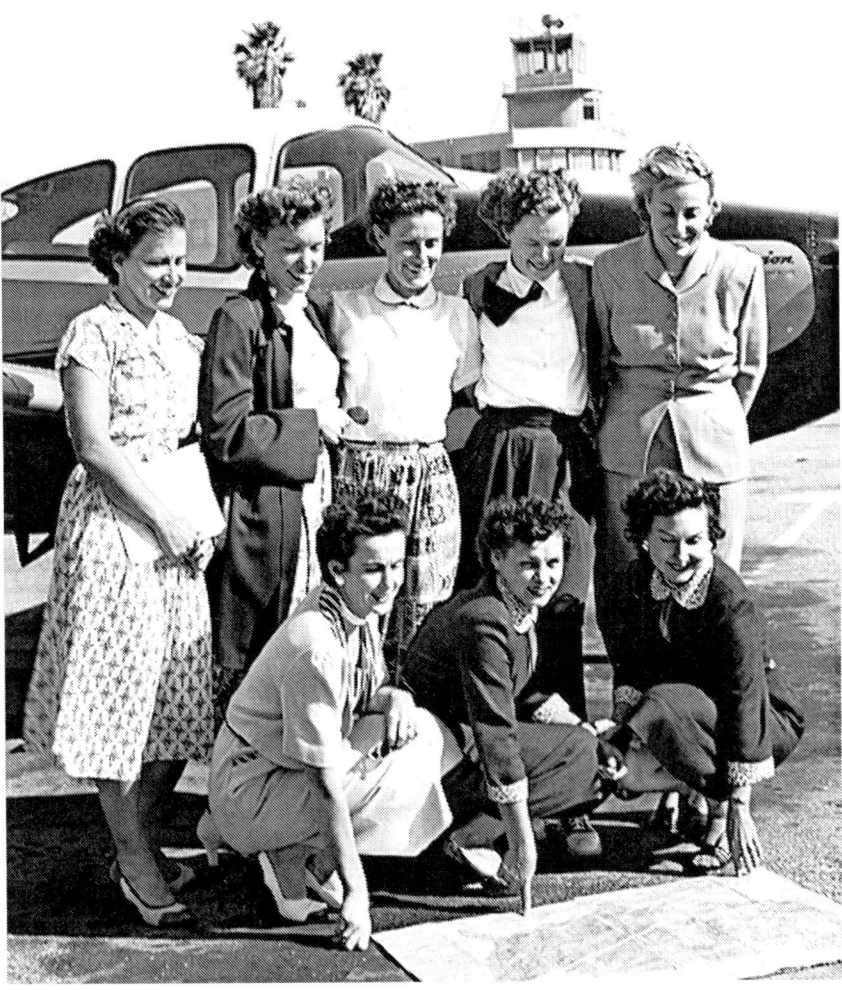

FRONT ROW, left to right: Barbara E. London, Frances Bera, and Marcella Duke. SECOND ROW: Jean Parker Rose, Donna Evans, Iris Critchell, Shirley Blocki Froyd and Betty Loufek.
Committee members for the 1955 Powder Puff Derby from Long Beach, California to Springfield, Massachusetts stopped to pose for one of Betty Loufek's air race publicity photos.
Photo by Wilkins of Long Beach, California

Chapter 8

I was attending a Ninety-Nines Chapter meeting one spring evening in 1959 when Iris Critchell said she had received a letter from a German woman, Elly Beinhorn, requesting an American co-pilot for the forthcoming Powder Puff Derby.

Elly Beinhorn had made long flights in the 1930s, and was one of the top three women pilots in the world at the time, quite famous in Europe. She was a fine navigator and pilot. One of her early long-distance flights was one she made in short stages from Germany to the Far East in 1931.

In 1932 she made a solo flight from Germany to Australia. She and her small plane, a 80-horsepower Klemm monoplane, then traveled by ship to South America, where she flew it over the Andes. While she was not the first to do this, it was still a great flight in such a tiny craft. Later she circumnavigated the African continent.

Elly had logged about 4000 hours over the years, but only about 350 hours since 1954. She earned her living as a writer, lecturer, and commentator on television programs. She had an enthusiastic following in Germany.

Elly planned to use her Derby experiences in articles and in broadcasts. Good publicity was essential to her livelihood.

I understood that; it did me no harm, either. Sometimes I got a new student from an article in a newspaper or magazine.

Nancy Bird-Walton, an Australian pilot as famous in her country as Elly Beinhorn was in Germany, had raced in the 1958 Powder Puff with Southern Californian ex-WASP, veteran flight instructor and experienced Derby racer Iris Critchell as her co-pilot. They took fifth place.

Nancy Bird was disturbed by comments from some of the aviation buffs that the race had accomplished all that it had set out to do and there was nothing more. She was anxious to have the race develop an international aura, giving the world an appreciation of America and of the accomplishments of women pilots.

Nancy visited Elly in Germany and talked to her about racing in the next Derby. Elly contacted various resources and managed to promote a Piper Comanche 180 from a dealer in New York. A German aero club donated money for expenses and the United

States Air Force agreed to fly her on a military dependents flight from Europe to New York.

Elly needed an American co-pilot because the FCC (Federal Communications Commission) at that time required American citizenship for a radio license. She could fly the airplane while in the United States, but was not allowed to use the radio. She had requested Iris Critchell, but Critchell was flying with her friend Barbara Thisted.

I hadn't raced since 1953 and thought it might be good to get back into it as a co-pilot for someone. Still, it was against my better judgment when I agreed to fly co-pilot to Elly. I knew there would be problems. It is hard enough to race with someone you do not know, but to fly with someone from a different culture and with different customs makes it difficult indeed.

A team needs to be friends in such a long race. Competition teams must trust each other and be cooperative. Elly and I would not even be thinking in the same language.

In 1959, 66 planes carrying 129 pilots were racing. They'd fly against prevailing winds on the east to west route from Lawrence, Massachusetts, to Spokane, Washington, a distance of 2470 statute miles.

The race was across the upper tier of states: Massachusetts, New York, Ohio, Indiana, Illinois, Minnesota, North Dakota, Montana, to Spokane, Washington. The previous year the race flew the lower tier of states from San Diego, California, through Arizona, Texas, Mississippi, Alabama, and Georgia, to Charleston, South Carolina.

Elly Beinhorn arrived in New York on schedule. I was to fly to New York to meet her. We'd pick up the Comanche being loaned to her by the Piper dealer in New York and fly it to Lawrence.

On the June day when all the schools and colleges in Southern California released their hordes, I took an airliner to New York.

It was difficult to get a flight and I had to take one flight to Chicago and transfer to a tiring "milk run" to New York.

Finally we arrived. I put the baggage in a cab and went to the hotel Elly had picked. It was not much of a hotel; pretty drab. Obviously she did not intend to waste the aero club's money. I called Elly the next morning and met her for the first time. We got on famously.

I arrived in New York early as I was told she wanted to fly the race route first, get familiar with it, then fly the race. This meant a round-trip flight of 5000 miles just before the race itself, an exhausting program, both financially and physically. Happily, she changed her mind and wanted to spend the time shopping in New York. I detest shopping but went along.

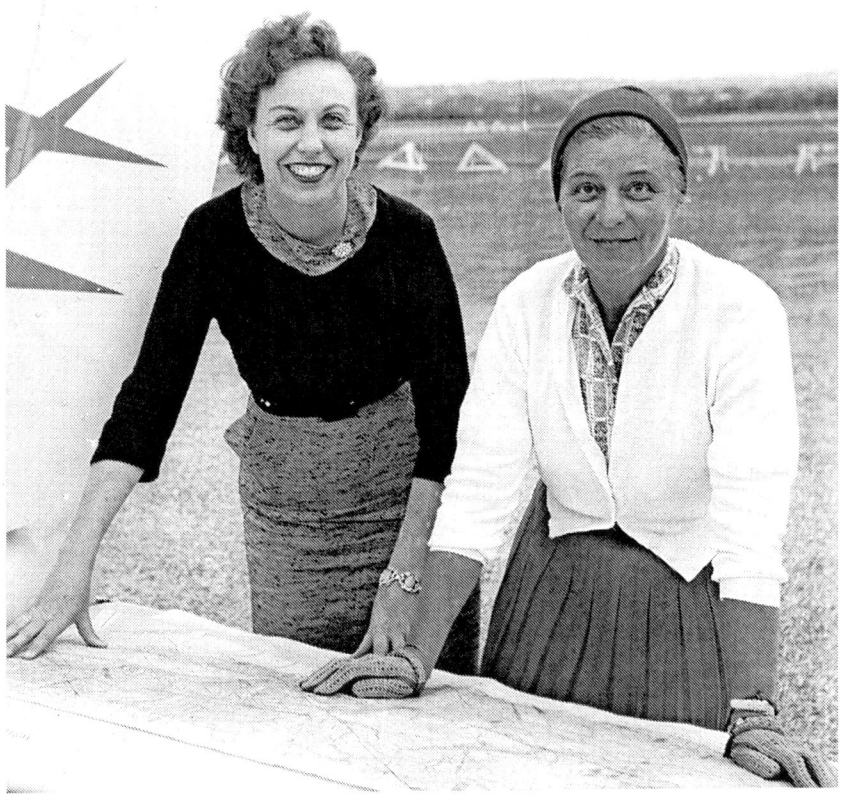

Elly Beinhorn, first German pilot to enter a Powder Puff Derby, needed an American co-pilot to handle radio communications. Claire volunteered. Race was from Lawrence, Massachusetts to Spokane, Washington.

We met the Piper dealer at his office on Wall Street. He talked to me as though I were the pilot-in-command. I knew why: Elly was the foreigner with the German accent; I was like him. If we had brought a man along to the conference, the dealer would have talked to him, even if he were not a pilot.

Elly said nothing in front of him, concentrating on the performance characteristics of the Comanche, its power settings, and other technical data. I was not familiar with the Comanche at that time. We did not use them at the school, but I knew that I would be checked out in it. We listened to all of the dealer's suggestions for achieving maximum performance.

As soon as we left the office, Elly made it quite plain that she was the pilot of that airplane. I did not dispute it; I had never assumed anything else. My job was to assist the pilot.

We went to the airport and flew the Piper Comanche to Norwood, Massachusetts, to visit Max Conrad. He and Elly were friends. We had lunch and then listened to the recordings of songs he wrote as he flew across the oceans delivering aircraft.

At Norwood I checked out in the Comanche, making a couple of landings. Flying it was easy; it is a nice craft.

From there we flew to Lawrence. Elly was still a fine pilot and navigator; no problem there. I began to teach her how to fly the race. In 1959 we still used a time clock set on a table with timing officials sitting behind it. If the time clock was placed near the beginning of the runway, we were to land as short as possible, turn at the first taxiway, taxi back near the time clock and stop the engine. After the propeller stopped whirling, the co-pilot climbed out of the plane and ran to the time clock. This was exciting and kind of dangerous; other plane crews would be heading for that same spot.

If the time clock was far down the runway, we should fly down the runway, and as we neared a taxiway, touch wheels to land, turn off and get to the time clock as fast as possible. The point was not to waste any time on the ground.

Until the time clock punched the page, our time ran as though we were still in the air. Racing strategy was planned with this in mind. Often only a few seconds separated the top winning times at the end of a race across the entire continent. The loss of seconds might be disastrous. Once when Fran Bera took second place, she was only 34 seconds behind first place.

Finally, in 1966, to the relief of everyone — race pilots, AWTAR Board, airport officials, and tower operators, timing lines were established on runways. The National Aeronautic Association appointed timers to be in the towers and the racers flew across a line for timing. Landing came later at the pilots' leisure and as air traffic permitted.

The timing change was made by AWTAR Board and airport officials in 1966 because they knew that the forthcoming races were to be heavy with entrants. In 1959 we had 66; in 1960, 85; in 1961, 101 planes.

The number of entrants for the 1962 and 1963 races went down because of the AWTAR Board's decision to upgrade pilot qualifications. The leading reason for that was the great difficulty presented in handling 101 airplanes. It was felt that 50 planes was about right and the only fair method to screen applicants was to raise qualifications.

Safety was always a major consideration to the AWTAR Board of Directors. The newer planes being flown were faster and heavier, with more complex instruments and radio navigation aids. There were more stringent procedures added by the government entities at and around airports.

When wind direction and velocity indicated it, pilots flew above 10,000 feet now that their airplanes were able to do that efficiently. To fly extended periods above that altitude can cause oxygen deprivation for most people, so oxygen equipment was being added.

Pilots-in-command up to 1961 could race with basic private certificates and their co-pilots or passengers did not need ratings. From 1962, the pilots-in-command must hold a commercial certificate or a private with instrument, or instructor ratings, with private certificates for co-pilots. This decision was applauded by the FAA and other aviation organizations. The race was leading private flying in all the great ways, as usual, and as planned by the all-woman AWTAR Race Board.

Because of the pilot certificate upgrading, the 1962 race had 54 planes and the 1963 race had 47. Race pilots are a hardy group and they are not to be denied. They upgraded their certificates and in 1964 there were 61 aircraft; in 1965, 79 entered; in 1966, 91, which caused the timing change to the flyby.

In 1970, 98 planes carried 181 pilots. For the 25th anniversary race in 1971, the race went from Calgary, Canada, to Baton Rouge, Louisiana, with 150 aircraft and 288 pilots. The last, the 30th commemorative flight in 1977, had 150 aircraft with 331 pilots and passengers.

Nancy Bird was successful in her prodding of foreign fliers. In 1959, besides Elly Beinhorn from Germany, Winifred Willmore of Sydney, Australia, was co-pilot for Clara Aldrich, and Adriana Fyn of Amsterdam, Holland, flew co-pilot for Katherine Palmer.

In 1955, we had had our first non-American citizens enter, a Canadian team, Felicity McKendry and Dorothy Rungling, but Canadians are our close neighbors.

Over the following years, racing pilots came from Japan, England, Germany, Bahamas, South Africa, Mexico, Philippines, India, Italy, Holland, with large numbers from Canada and Australia.

Nancy has accomplished many wonderful projects over the years. Her autobiography, *My God! It's a Woman* tells of her early years as a pilot in Australia. In 1966 she received the Order of the British Empire from Her Majesty the Queen at Buckingham Palace; In 1987, the degree of Master of Engineering (honoris causa) by the

Sydney University; in 1990, the Order of Australia. Carry on, dear Nancy Bird.

Elly and I practiced airport patterns with an imaginary timing table at the side of a taxiway along with some of the "upstairs" work. Then she began to show resentment. She began to ask other people how to fly the race. Apparently she had gotten together with some German male friends and they had told her how they would fly the race.

They advised her to throttle back, cruise across the United States so as not to have to land at every stop. They had never raced in a cross-country, did not know the rules, never talked to anyone who had raced the PPD. It is hard for many people just to say, "I don't know."

I told Elly that we could not cruise across the U.S. Our handicap was based on cruise and we had to do better than that or we would have a minus score. We must go full throttle from stop to stop and try to get on the plus side. Little problems began to grow but in general we were pretty good friends. She taught me a few German words and we got along fairly well.

In New York we had separate rooms. When we arrived in Lawrence they gave us separate rooms, but with a connecting bathroom. Elly told me immediately that she did not like this but that she would give it a try. The first night she found it convenient because she could dash into my room whenever she pleased to tell me whatever was on her mind.

Then Elly told me she was to be the first into the bathroom. Well, all right; I did not care. It seemed a little odd for her to make this particular statement. Whenever I traveled with others we split the time with no problems. Maybe she worried I would use all the towels...or all the hot water.

Elly was quite positive on this point: she was to have the bathroom first. I assumed she would try to wake up the same time as I or make sure it was earlier; that would be fair, right? Surely she realized that even though I was at least 20 years younger, I still would have to go sometime.

Well, the next morning she either stayed in bed quite a long time or else I woke too early. Finally I had to use the bathroom. Then I went on my way. Later that day she came to the field and jumped me about using the bathroom first. I told her I had waited as long as I could. The next day I woke up earlier or she stayed even later in bed, because finally I had to get up and use the little girls' room. Then I went on about my business.

Elly caught up with me at the airport and let me know in no uncertain terms that I had not done as she had directed.

"You used the bathroom first!"

"Well, for cris' sake!"

Our cordiality began to deteriorate.

Elly began to study the maps and plan her race. She did not want nor need my help. I wandered around to see some of my old friends who were also racing. They knew how she was treating me and asked how I could take it, encouraging me to answer back. But, I was not snapping back; I was trying to go along with the deal.

I knew Elly was nervous, anxious about making a good showing. She was the first German to fly the race; she had to return home and face her friends and fans. Here in America she faced news reporters from American and German papers as well as possible rejection by some because she had been a friend of Hitler.

I had a car at my disposal. I got up early the next morning, quiet as a mouse, dressed and left for the field, leaving the place to her.

I came back to my room about noon to find my things removed.

I was thrown out because I had used the bathroom first. Elly could have moved herself, as another might do. This was a power move to show me she was in charge.

Some of my friends asked me if I wanted to fly with them. But this would ruin any chance they had of winning. The rules did not allow anyone to be added after a certain time, but someone could be dropped off. I told them, "No, I'll try to stick it out."

Fran Bera was flying with her sister Edna Bower. Fran had won the 1958 race, but in the final standings would be in 16th place this time. Fran and Edna had also won the 1955 race. Their sister, Helen Albani, flew with Fran to second place in the 1954 race.

We had a good time teasing each other, enjoying the day while trying to ignore the butterflies, the anxieties, that arise in all competitors everywhere as the hour nears for the flag to drop or the starter's gun to bang.

Joining us in our camaraderie were Pauline Glasson and co-pilot Ardath McCreery (they would place fifth), Iris Critchell and Barbara Thisted (ending in sixth place), and a dozen others of these wonderful ladies.

All wore most attractive summer dresses, some designed and sewn by the wearers. Many of the teams wore identical outfits from their own desire to do so. One of the teams even wore the same colors as their plane.

Pauline Glasson, when the final race was run in 1977, had flown in 24 PPDs. A flight instructor, she made a point of taking one of her own students with a new private license in hand as her co-pilot, whenever that was possible. She held the record of having the greatest number of different co-pilots. She continued this policy when the Air Race Classic began after the Powder Puff Derby had ceased.

Over the many years Glasson eventually instructed the grandchildren of some of her original students. She has accomplished a great deal for aviation in many ways beyond instructing.

Pauline Glasson is just one example of the millions of unselfish women who multiply by many times their presence, their influence on their own worlds.

It was the morning of July the Fourth. The moment had come. The official flagman shook out his flag and stood ready, his radio communicator waiting behind him. The first dozen race plane engines were started.

The timing runner stood at the table with the first plane's logbook in her hand. At a signal, the logbook's first page was punched, the runner ran to the lead plane and handed the book to the pilot. The flagman dropped the flag. The pilot pushed the throttle forward and the plane moved smoothly down the runway, gathered speed and rose. Every twenty seconds another plane was flagged off.

Our turn came and as we climbed on out I got back into the spirit of the race. We each had a set of charts and as we flew I tried to tell Elly where we were, naming the towns.

She answered, "I know where we are. I will do the navigating!"

I answered, "Okay!", threw my charts into the back seat, slumped down in my seat, and closed my eyes to get some sleep. I needed to use the radio only as we approached our first timed, refueling stop, 248 miles from Lawrence.

We neared our first stop, Broome County Airport at Binghamton, New York. My competitive spirit rose again. As we entered the final approach to the field, I said, "Now we should fly as far down the runway as possible as the time clock is at the far end of the field. It's a long runway. Just before we get to that last taxiway, we set it down, turn off and get to the time clock."

As we approached the runway, Elly began throttling back and slowing it down. I said, "No, Elly, fly on down the runway as far as you can, then throttle back."

She kept throttling back, throttling back, while I kept saying, "No, Elly, keep it flying down the runway." She throttled all the way back and we landed right at the beginning of the runway. It was 5000 feet to the timing clock.

I was stunned and did not say anything more. We taxied and taxied and taxied and taxied....Finally, we got to the time clock. Elly stopped the engine. I climbed out and ran like a fool for the time clock. A waste of energy as we had lost all our time taxiing. We lost any possibility of placing right there at the first stop.

I walked back to the airplane, tossed my logbook into the plane, got my lunchbox out, walked about a quarter of a mile down the line of parked planes, sat down under a wing and started to eat my lunch. A chest pain hit me; I thought it was a heart attack. It turned out to be chest pains from tension. I was ill for about an hour and thought I might be unable to continue the race. Since that time, whenever tension reaches a certain point, I have this pain.

Eventually I walked back to the airplane. I will never forget the sight of Elly standing on the wing, hands on her hips, saying frostily in her German manner, "And vere haf you been?"

"Well," I answered, "This is it. I don't think I will go on. I'm trying to help you and you are fighting me."

"We started out as partners and we're going to end that way."

"I'm not so sure. Things are going to have to change if we are going to do this. I'm sick."

We had to stay in the same motel that night, but whenever we could I stayed in a different motel. My friends were aware of what was going on and were concerned. They wanted me to fly in their airplanes for the rest of the route. We really could not do that, so I turned them down.

Ahead of us were Margaret Callaway and Helen Herd, Margaret's niece, a student pilot. Ahead of them were Margaret's daughters, Margo and Sandra. Teenagers, they were the youngest ever to fly the race; each had several hundred hours on private licenses. They were taught to fly by Margaret, who was a fine flight instructor. The daughters took third place, obeying their mother's words but not doing as she did, as a news item reported.

Margaret and Helen were disqualified for remaining overnight at an undesignated stop. Perhaps weather or lack of fuel caught them; sunset would come before they could get to the next airport with a time clock.

These are the breaks of the game. During the race, six other planes were disqualified for the same reason. An unbreakable rule was that flying was only to be done during daylight hours and under FAA visual flight rules.

Margaret and Helen continued their flight to Spokane the next day to join in the fun, already planning to try again the next year. This was a flying family; Margaret's husband was Lt. Colonel Richard

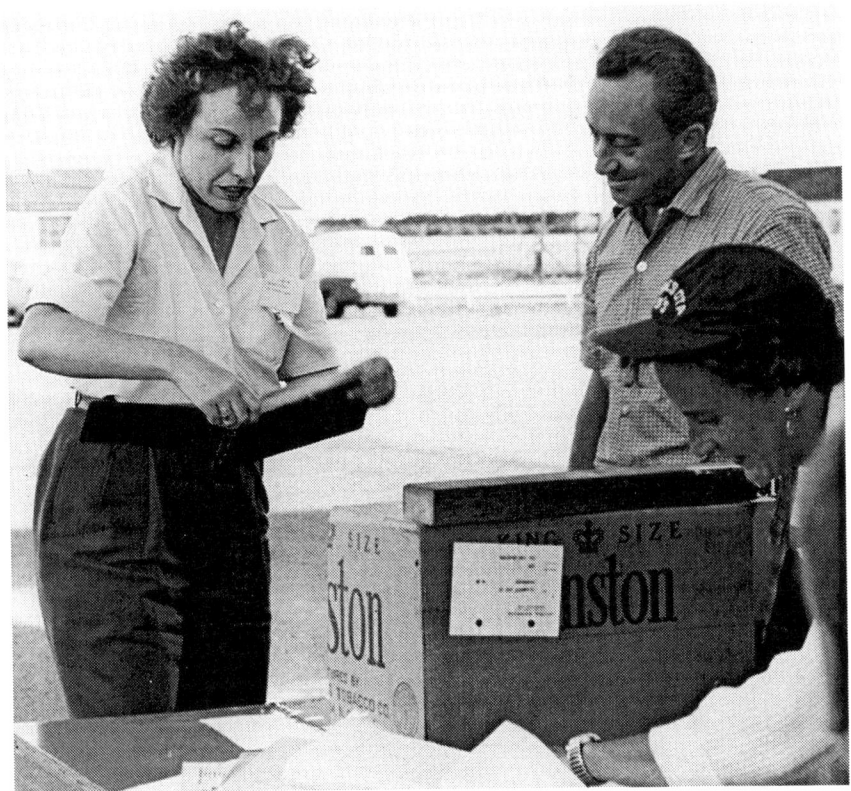

Claire at one of the timed stops, trying to get the page punched so race time would be stopped and plane refueled. (1959)

Callaway, an Air Force senior pilot and a deputy commander of an air reserve training wing at the time.

At sunset, race planes were scattered from Chicago, Illinois, to Helena, Montana. One plane crew, Margaret Ringenberg and Lois Laymon, flying a Beechcraft Bonanza, wanted to reach Spokane that first day, but were slowed by head winds and having to navigate around storm clouds. They reached Helena, Montana, at the end of the day and stayed overnight. They went on to the terminus the next morning.

They ended in 30th place, but they did not care.

Reaching the terminus first has its own rewards. Everyone is excited, the press takes a lot of pictures and sends them everywhere, gifts are presented, and the crew does not have to worry anymore about the weather. Not the least of the rewards was the joy of sleeping in the next morning instead of getting up at three o'clock.

Fargo, South Dakota, was a must stop. A squall line lay to the east of it, across the flight path of the race pilots. We were able to wend our way between thunderheads, even though our small planes bounced around like ping pong balls.

Once away from the coast with its morning fog, most of the race pilots got up around three o'clock and prepared to leave at the first moment they could punch the time clock after official sunrise. Summer thunderclouds and squall lines get vicious in the afternoon. Most of the pilots planned to end flying for the day by noon.

Weather knowledge was essential. Should a team race on to the next stop before the squall line closed the airport, or will the winds be better the next day if they wait where they are? If they decide to race ahead but know they will reach the next airport just before official sunset, will other planes be in the pattern, delaying the team until the time clock clicks over to the minute after sunset and they are disqualified? The race ended at noon, July 8, so all had three full days.

Fran Bera and Edna Bower's plane had its electrical system fail near Three Forks, Montana (not a timing stop). They landed, got the system working and took off only to find they could not raise the landing gear completely. This meant they could not operate the plane at normal speed. They reached Helena and got it fixed. The race would end at noon Wednesday, and that morning Fran and Edna took off. A cell cracked in the battery, located on Edna's side, splashing acid on her tennis shoes.

"What's that smell?" asked Fran as she opened her window to let out the fumes.

"Whatever it is, it's eating holes in my shoes," said Edna.

They returned to Helena and punched the time clock once again. Mechanics replaced the battery with a smaller one; it would not start. It was recharged; the race team punched the time clock and took off. While on the final leg into Spokane the radio went out. They still placed 16th. It is obvious that without the problems they would have done extremely well.

In an interview with the press, Fran shrugged, smiled, and said, "I feel you just have the breaks some years; other years you don't. You can't win them all. Next year will be another race, and another chance."

In Helena, Montana, Gerry Mickelsen, Ruth Wagner and I shared a room. They knew my situation and wanted me to stay with them. All along the way I never had to stay by myself if I did not want to, and I had company at meals. My friends were great. They made the situation more acceptable.

When we arrived in Helena, Elly's German self showed up again. She wanted to go to a different hotel, of course, and she told me that I was to call her at a certain hour.

The girls handling that route stop said, "Elly, Claire doesn't have to call you. We'll call you."

She turned to them and said, "Claire will call me."

They were taken aback. They had never before observed this sort of thing going on. I said nothing.

The press were there to interview Elly. She did a good job and was gracious to the press.

The next morning I got up and went out to the field with Ruth Wagner and Gerry Mickelsen to get the weather report and see if we could go on. I intended to call Elly when the time arrived.

Elly showed up, indignant that I had not called her. I looked at my watch; it was not yet the time she said to call.

Helena was our last timed stop before the terminus at Felts Field, Spokane, Washington. Ahead was mountainous terrain. Weather reports noted scattered thunderstorms over the mountains. It seemed poor to me and I did not want to go. On the other hand, I could not stand another day of the cockpit situation and thought how nice it would be to get the trip completed.

Elly was scheduled for a press conference in Spokane. The publicity was important to her. She wanted to go on, too.

According to the current weather report, the tops of the clouds were at about 10,000 feet above sea level.

We climbed into the airplane and went through the checklist. Everything was all right. I ran to the time clock and punched us out. We climbed on out of Helena Airport and headed for Spokane, going over MacDonald Pass. Its floor was at 6300 feet, which meant the mountains on each side filled those clouds with rocks.

We got to the pass to find thunderstorms where we expected scattered clouds. Further, the cumulonimbus clouds were developing fast and moving together. Some were 20,000 feet high. I saw open spaces between the clouds where we could go over the top at just a little above 10,000 feet.

"Elly, we can easily go over the tops of the clouds through those canyons between the thunderclouds. It should be clear after that."

She answered, "Well, we cannot go over 10,000 feet without oxygen."

"Oh, the heck we can't! There's no problem. There's no reason why we can't go higher than 10,000 feet without oxygen." We would be there only about 20 minutes. I have hiked higher than that for hours.

"We will not go higher than 10,000 feet without oxygen!"

"Oh, all right!"

I called the flight service on our end of the pass as we neared it and asked how it was now. The observer at the station said, "Fine, fine. It's clear, visibility such and such...."

I said, "I don't know what side of the building you're looking out of. Have you tried your back door?" Because on that side the sun was not shining, the sky was heavy with clouds that looked as though they were dumping snow, hail, and rain.

The flight service operator said, "Oh, you can get through the pass easily. It's the usual route used by pilots here."

I said, "Well, we'll give it a try. We'll head down towards this pass and see what it looks like."

According to the map, there is a big curve in MacDonald Pass. We had to know exactly where this curve was and plan ahead for it as we entered the pass. I measured this first section carefully on the flight chart and translated that into minutes of flight.

I called the flight service closest to the other end of the pass and asked how the weather was there. They said it was fine.

I called our closest station again and said, "It looks pretty bad in there. It has heavy storms and the visibility is down."

"Oh," he said over the radio, "You can make it. I know you can. After all, a Bonanza went in there just a little while ago and I have heard nothing from him."

"Oh, good, good! That reassures me tremendously!" I turned to Elly. "This is for the birds!"

Elly answered, "Oh, we can make it through the pass."

We entered the pass. The ceiling (cloud level above us) dropped sharply. The clouds, combined with heavy rain, made visibility nearly zero.

I said, "The only way we'll make it is in a rubber sack. I'm for turning around and going back. Come on, Elly, turn."

Elly started the turn but it was not tight enough. The canyon wall loomed right in front of us. I grabbed the wheel — the only time I took the plane away from Elly — tightened the turn and missed the canyon wall, rolled out and headed back to Helena.

The Bonanza flashed by underneath us on its way back. Its pilot called us on his radio and said he had heard the transmission from the flight service station and tried to call us and tell us not to do it, that he could not make it through.

We returned to Helena and punched the time card. This was one hour and 20 minutes of lost time, as far as the race was concerned. After punching the time clock I told the officials there was trouble in the cockpit. At this point we were not speaking. Later that afternoon the storm cleared and Elly wanted to leave but I refused. We made it into Spokane the next morning through scattered puff clouds.

I shudder sometimes, thinking how close we came to being dead on the side of a mountain. I am devout about surviving.

Most of the racers had arrived and were relaxing among their friends. There is always fun at the end of a race. Pressure is off, poolside sitting is in with tall-tale time, and renewal of friendships. Aileen Saunders and Jerelyn Cassell had flown a Cessna 172 to first place, which they'd learn the next day. Juanita Newell and Jimmye Lou Shelton had flown a Piper Comanche to second place. The teenage Callaway sisters, third.

Elly wanted me to meet friends of hers who were with a German Consulate friend. I elected not to go meet her friends, but to just stay with my friends. At that point I wanted nothing more to do with any German, good, bad, or indifferent.

Since those days...well, even that summer, on a three-week trip to Europe, I visited Elly in Germany, and in 1967 she was out here at the end of the race at Torrance (California) and again in 1991 I saw her at Santa Monica Airport and we exchanged pleasantries.

The experience taught me a good lesson, or should have. I flew the Angel Derby with someone I did not know, and it was a difficult trip, again. Now I refuse to consider racing with someone I do not know.

When I am going to fly a race with a gal, I find that it is a good idea for us to attend a Ninety-Nines Sectional or a fly-in someplace or a weekend with a group of people, just to see if we can get along. If we do not like the same things, if I irritate her or she irritates me, there is no point in going on a two or three week trip together.

Before that and since then I have been lucky with my co-pilots so far as getting along is concerned. As for Elly, it was her way of doing things. She had flown solo most of her flying life and did not know how to deal with a co-pilot, especially a foreign one like me, an American woman.

Chapter 9

I started my business in June 1960, with $3500. I put $3000 down on a plane, a down payment on the rent, a down payment on the insurance, and opened the Claire Walters Flight Academy.

The little airplane I bought was a Cessna 150. I leased another Cessna 150 and a four-place Cessna 172.

New students and many students from my last employment came and the business began to grow. I did all of the instruction, of course. When golf champion Betty Hicks and I raced in the Powder Puff Derby right after I started the school, an instructor working for Cloverleaf Aviation and who also worked for me part-time, took over while we were away.

Then Emma McGuire walked into my office. At that time, her husband owned The Fireside, a market in Santa Monica, as well as others around the country. She worked in the Santa Monica store.

A number of Emma's women friends flew. They'd stop at her market to buy groceries. Emma would ask about their flying. She was intrigued with the places they were flying to for lunch or just flying to for fun. Her five children were grown; she didn't have to hurry home for them.

Emma began to think that maybe this was something she ought to try. She went to the airport and started taking lessons from another school, but they were not good to her. I had started my school and some friends of hers told her she should take lessons from me. She did and we became good friends; we have been friends now nearly 40 years.

Emma is the type of person who is willing to do almost anything and is always gung ho. But she is careful to look beyond the task and be sure she knows what she is doing.

Emma soloed a Cessna 150. Her husband bought her a Cessna 172 and we taught her to fly that. She loved the four-place plane and whenever she had an hour or two she'd go flying locally.

Emma seemed to be having some difficulty getting away for her solo cross-country flight. In the natural course of things toward getting a private license, students are taken on dual cross-countries and then sent alone on a cross-country flight when the instructor says each is ready.

She must do the solo before she could take the next step towards her private license. Once she had the license, she could take her non-pilot friends flying.

My flying school was growing; we needed another Cessna 150. The only one we found available was a new one in Santa Fe, New Mexico. We made arrangements to buy it and prepared to go to Santa Fe to pick it up.

Emma said, "Let me take you girls to Santa Fe."

"All right, Emma."

At Santa Fe, when we got ready to leave, the other girl and I got into the Cessna 150 together. You see, I had this plan....

"What are you girls doing? One of you flies with me."

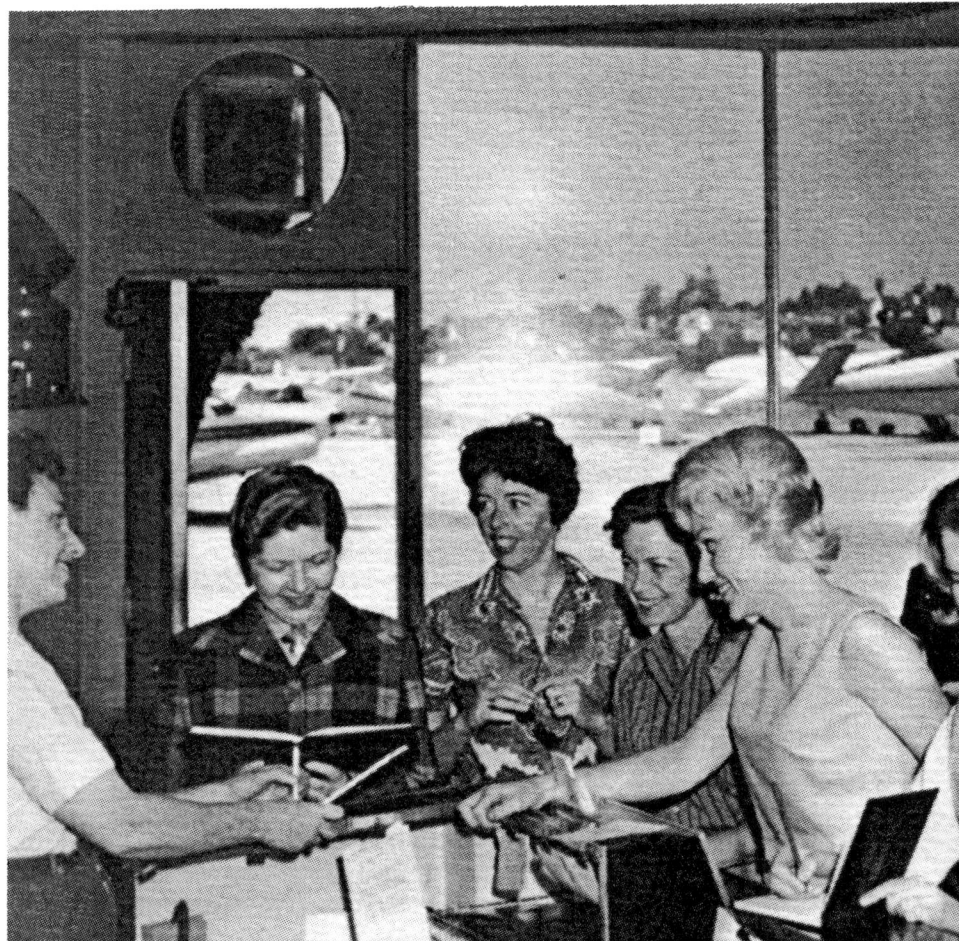

Claire Walters Flight Academy opens in June 1960! Students and flying friends soon fill and excellent equipment to fly. Claire (far left) is helping them with their logbooks.

I said, "No, Emma, this is where you start your solo cross-country."

It would have been more fun, and more comfortable, to take turns flying the 150 and being a passenger in the 172, but I believed this was a real opportunity to get Emma signed off.

We were going to make her fly back alone, but the way we had it figured, she'd not really be alone. We'd fly along within a mile or so of her and keep in contact on the radio. Emma was upset with us, a bit shocked. It was a long way to fly alone, no one to talk to. She told my sister recently that what really worried her was my description of the New Mexico and Arizona thunderstorms and how dangerous they were.

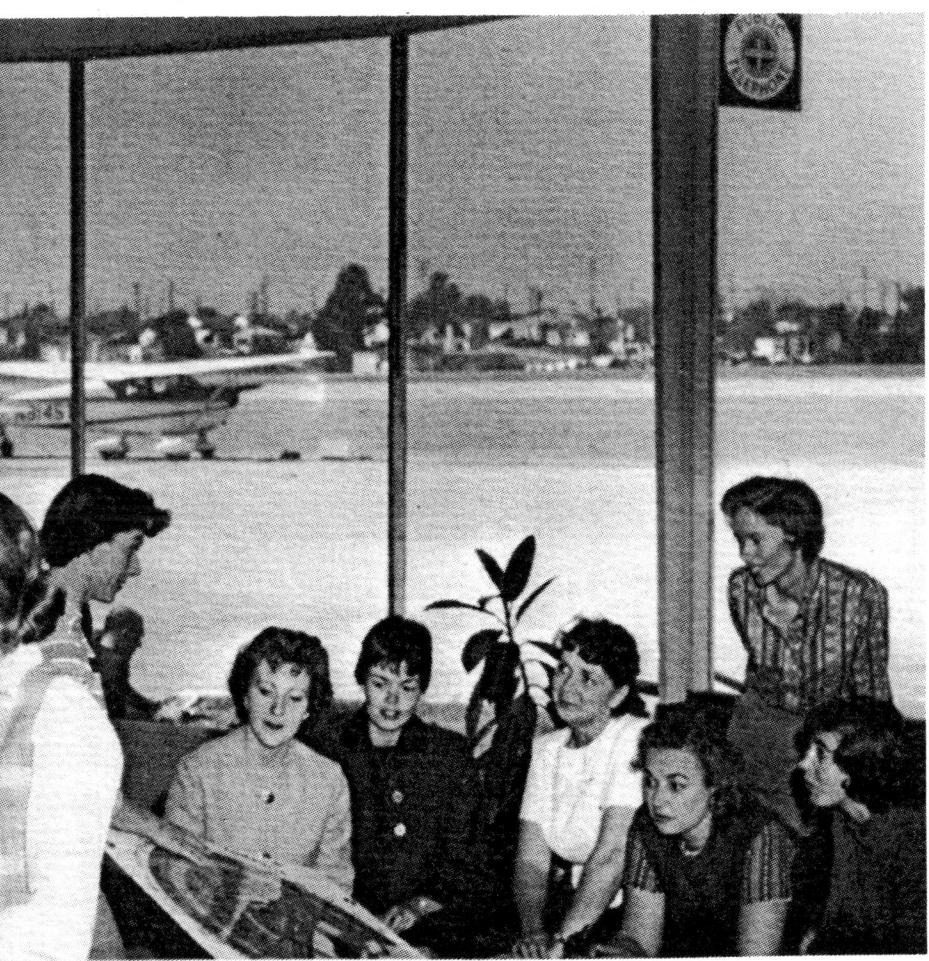

office and it became popular for those seeking courteous treatment, fine instruction,

We all agreed that our RON (remain over night) was Winslow, Arizona. We'd refuel enroute at Holbrook, Arizona.

We tried to keep in touch by radio as we flew. We saw the glint of her airplane right behind us much of the way, then we lost sight of her. Over the radio we asked her where she was; she answered that she was inside of a rainstorm. Right next to us was a rainstorm and we thought she was probably right there. We weren't too worried, except that we couldn't see her.

We reached Holbrook, refueled our plane and waited. No Emma. We checked over our plane and waited. No Emma.

I began to feel sick. I was just beside myself. Here I'd sent her out on her own like that, and while it should have been plenty safe — she'd had lots of flight experience — it was still her first solo cross-country.

We flew on to Winslow and tied the plane down for the night. Still no Emma. We went into Frontier Airlines and they sent out a teletype to all their stops to see if she had been seen any place. No, she hadn't. We checked with the weather bureaus and the flight services on our route. No Emma. Oh, migod!

At sundown, Emma taxies in and I became the happiest flight instructor in the country at that moment after being the most scared.

Seems she thought she might run a little low on fuel before reaching Holbrook and decided to land at a field before it. She found one but no one was there; they were all in town at a ball game and the pumps were locked. There was no phone or she'd have called Winslow and left a message for us.

She took off and flew to the next small airfield. Same thing. Finally, she made it to Holbrook, refueled, and flew to Winslow. The next day we had her continue her solo flight, but we made sure we stayed close to her.

This was one of those incidents you later look back on and laugh about, but at the time it was not funny. In fact, I can tell you that that was the most scared I have been in my entire instructor career.

Emma passed her private license flight test and went on to get her commercial, instructor, and instrument ratings. She instructed for me for years. She flew in several Powder Puff Derbies as well as in other races. While she never won, she had fun and gained much experience. She is loved and respected by all of us who know her. She is quite a gal. We flew in her Cessna 182 in the 1986 Air Race Classic.

The new Cessna 150 I had just bought needed a paint job. It was only partially painted; the rest was unpainted aluminum.

Here, near the seacoast, aluminum corrodes easily. It is necessary to paint the entire fuselage and wings, everything. If you leave it unpainted, you must buff it and it'll look nice for awhile, then you buff it and it looks nice for awhile, then pretty soon you buff through the alclad and you don't have anything to protect your aluminum. It is necessary to paint it right away before corroding begins.

I had spent almost all I had on the down payment for this second plane. I had $87 left and used it to buy paint. A friend and I brushed on the primer ourselves. It was a bad primer — some new kind — and it didn't go on easily. Finally, the plane was ready to paint.

We borrowed a hanger from a friend, plugged up all the holes we could find in it so we couldn't get paint on any other aircraft, then moved the plane into it.

We covered the windows and other parts that needed protection from the paint and those that weren't to be repainted, then her brother started painting. I had the plane fully scheduled for flight from seven the next morning through the entire day, so it was necessary to finish painting it that night. Fortunately, it was a dry night, which is not all that frequent in Santa Monica.

He had a spray gun. He did the best he could, but I guess it was a pretty rough job, judging from our friends' comments. We worked all night long, then while the paint was drying we left to have breakfast. When we came back we pulled off the paper protecting the windows and so on, and the plane was ready to fly by seven. Whew!

We were dog-tired, but the airplane was ready to go. I had saved several hundred dollars.

Over the years I bought many replacement aircraft for my school. It was important to keep well-maintained and fresh looking aircraft, and there were many such planes available for instruction purposes into the middle 1980's when the companies stopped manufacturing them mainly due to liability insurance problems. It took almost ten years to resolve these and then to start building new trainers in the middle 1990s.

Betty Faux and I took a TWA Constellation — it was the cheapest flight available — a "milk run" which stopped everywhere, to pick up another new plane. The flight attendants told the captain that we were going all the way to Wichita, Kansas.

Captain Charles M. Davis came back to see us. His first question: "Wasn't there anyway to go there but this?" We told him about the school, the new plane, and about being low on funds. He invited us up front to his office, the cockpit.

After we returned to the school, Charlie came to the school and I checked him out in a small plane. We have been friends ever since.

When Emma McGuire took the flight test for her private license it was with Pete Campbell, a brand-new FAA inspector on our field, and a friend of mine.

Just ten years later, in 1970, Campbell brought his number one team out of Oklahoma City (FAA headquarters) to give the lectures at the two-day flight instructor clinic which I chaired. It was extremely successful.

The clinic made about $1000 each for the Los Angeles and Long Beach Chapters of The 99s. We sponsored several more; other groups have sponsored many more.

These are very successful clinics. The FAA team lectures on fundamentals of learning, psychology, safety, instruments, analysis and performance of maneuvers, weather, instructor responsibilities, so forth.

Emma McGuire is quite a wonderful person. She has extreme patience. I've never seen her more than mildly angry about anything. Extremely easy to get along with, unselfish, always cooperative, willing to do anything and everything. She had a deep interest in her students, always ready to give her time to someone to get them ready for a license and help them prepare for the written exam.

I always used her for my ROTC (Reserve Officers' Training Corps) students. These were college students, extremely intelligent and goal-oriented. They had such a short time allotted on the program that they needed someone like Emma to take an interest in them, take the time, be willing to work with them and make them into good pilots. She even took them up in her own airplane occasionally. They just loved her, followed her around, gave her Christmas presents. They always recommended her to their friends who were starting to fly.

Emma's flight instructing was the model of teaching. She had endless patience and took great interest in her students and their progress. It took some doing to fit the student's hours into the plan, or the plan into what hours he had coming. The ROTC only gives 35 hours so it has to be a sure-fire program, with complete devotion to it.

Students are people, and quite often flight instructors forget this, forget that they don't like to be yelled at, forget that when one is tense and nervous it is difficult to work. Some give their students a bad time. When you yell at a student or bawl him out unnecessarily, he becomes nervous and tense and the performance he gives is less than desirable. One cannot think while one is angry; logic gives away to emotion.

It is important to treat your student right — not only so that he will remain your student and not seek other instructors, other flight schools, but becomes your devoted friend.

If the flight instructor wants to remain on his job and eat, he'd better make sure his students like him so they will tell others. Then he will eat regularly and be overrun with students.

I practiced patience all the years I instructed. I worked hard with the student to get whatever rating he was after, keeping track of his records so that everything came up when it was supposed to.

Students are normally shy, wanting to please and wanting to learn, trying to learn. The flight instructor who doesn't recognize this is foolish. He must understand that students are human beings first, and students, second. A human being doesn't have to be yelled at in order to understand a point.

Most of my students became my friends. They'd come back for additional ratings or send others to be taught to fly. Many over the years came by just to say "hello". These people and I liked and respected each other.

I helped to plan their work so it would have a continuity of pattern, a goal to work towards. A goal in life was something I took a great deal of interest in. As a result, some of my students are now captains on airlines, flight instructors, own flying schools, fly crop dusters, and of course, the many others who fly for pleasure.

It is possible to develop a friendship with the student — and I always tried to — and still have a profitable relationship for us both, developing a good student into a good pilot, and, in the case of the very young, into a good student.

I began my instructor career in 1946 in the Piper Cubs, the tandem-seats yellow birds with a tail wheel in which from the late 1930s through the '40s most students got their first eight hours of dual instruction, at least. But only about one out of eight could learn to land Cubs properly. The ability to judge your altitude above the ground and then maintain close control so as not to groundloop on landing is not easy to learn.

We instructed in Cessna 150s at my flight academy. This is a two-place, side-by-side, single high wing plane with a fixed tricycle landing gear; that is, the wheels do not retract into wells. Cessna Aircraft Company first brought out its tricycle gear planes in 1954.

Despite these easier planes to fly and to land, today student times for dual instruction before the first solo is up to an average of 12 hours. There are more instruments to watch and more radio work to do. With the massive air traffic, this is all to the good for safety's sake. And, as a consequence, the student landing accident rate is down.

In the Cubs, we spent much time practicing take offs and landings during those eight hours, and the final hour before soloing was usually spent this way.

When I was reasonably confident that the solo would be successful; that is, the student could get the plane off the ground, climb to the proper altitude, follow the pattern around the field, and come in and land all right, I'd have the student taxi to the edge of the landing strip. I'd climb out and wave him on his way.

I was always reasonably confident he could do it or I wouldn't have climbed out. But, humans being what they are, one can never be sure of what might happen once the student is out there by himself...and he realizes he is out there by himself.

When I was a brand-new instructor, I worried hard through the first few solos. This mother hen was letting her lone chick out of her control for the first time. I'd watch the plane all the time it was in sight and pray as I walked up and down at the edge of the runway.

Almost always everything would go all right and a grinning student would taxi up for the solo initiation ceremony. But sometimes a student "chickened out" at the last moment when coming in for a landing, and go around again while my heart sank.

A good deal of pressure was off the instructor when tricycle gear planes became available for instruction work. We could let the student do almost everything from the start, even do the landings without a great deal of help, getting the feel of that at once.

Instructors know the solo will be successful. No matter how bad the wind, the landing will be "good", or it can be made to look good even if it is really poor.

Tricycle gear planes lose five miles an hour air speed against their tail wheel counterparts, but the planes will not groundloop and the pilot's visibility is much improved during the ground run and while taxiing on busy airports.

The number of students who can learn to fly successfully is greatly multiplied; it is not limited to those who have perfect control over the landing.

Of course, sometimes this easy handling can backfire by raising the confidence of the student too high, and he thinks he is ready to solo in minimum time (eight hours dual). Like the incident I can remember now with amusement, mixed with shudders from thinking about what might have happened.

Normally I was on time with my students. Oh...maybe just a little bit late, sometimes...just enough to let them get the plane checked out so we didn't waste any time.

On this day I am in the office when one of my students comes in. He picks up his board and the keys to the plane.

Me: "I'll be out in just a few minutes."

He: "Oh, no, you don't need to come this time. I've had eight hours of dual; I can go solo now."

I've been forever grateful that I was there.

Another student, a gal with three-and-a-half flying hours, came in and demanded a key to an airplane. She wasn't scheduled. I told here I could take her later. She said that she didn't need me; she'd dreamed she'd soloed — so she knew she could do it.

In 1975, we had a strange student attempting to learn to fly. I was determined to never let him solo because of his mental instability. Because I was afraid someone else might solo him, I flew with him.

One day I had to go on a student cross-country dual so I asked Emma McGuire to instruct him that day. I told her that he was weird but not dangerous. I couldn't have been more wrong.

They taxied out to the runway in the Cessna 150. She took his hand and put it on the throttle, where it was supposed to be on take off, as they were climbing out. He accused Emma of sexually molesting him; of raping him. This 6'2" 180-pound, 28-year-old man was accusing Emma, 5'3", tiny frame, of raping him.

They flew the pattern, landed, and he ordered her out of the plane. She refused and ordered him out. She taxied to the parking space while he ran to the office. There, he stood over my bookkeeper, Gitta, threatening her and one of my tall male instructors with a fingernail file. Then he stood over my two-year-old granddaughter Michelle in her playpen and told my daughter Susan (she worked for me in the office) that he was going to kill the whole family.

Susan told him that he wouldn't be saying these things if I were there — then seeing me taxiing in, said, "Here she comes now." He took off running.

He phoned me twice that week to threaten me with death. I called the police, of course.

The fellow enrolled in several other schools, causing havoc until his medical was rescinded — that's all that anyone would do.

Right after the incident with Emma, who was in shock, a lot of her former students came running in to demand from Emma what was wrong with them since she had never tried anything with them!

A fellow waiting to fly with an instructor walked up to the counter, and with a twinkle in his eye, said, "I want Emma!"

It took a long time for a shaken Emma to recover from this experience. One day while I was telling the story, Emma quietly said, "But, it only took six minutes to fly the pattern, Claire."

I said, "Emma, I have it on good authority from several madams that the average male takes only six minutes!"

She laughed and began to show signs of recovering from this thoroughly devastating experience.

In the summer of 1967, I was approached by Steve who asked me to flight-check him; that is, ride with him on the flight test part of the private license. This is done after one has passed the written examination consisting of questions and problems on navigation, flight rules, so forth.

He said he'd taken a couple of check rides (later I found there had been several), that he had flunked them, but that he wanted a flight check for his private license in his own airplane, an Aeronca Champion. Because of bad weather, it was a couple of weeks before we could get around to it.

In the meantime, two men came to my office and asked me if I could tell them what an Aeronca Champion looked like. I said, yes, that I had a plane built along the same lines. I took them out to a similar airplane with a 65-horsepower engine. It is a high-wing plane with a tail wheel and tandem seats.

The men were from a TV studio; they were making a Lassie TV movie and needed an Aeronca Champion for the studio shots. The flight scenes were already filmed; the same type of aircraft must be found for the studio scenes. They discussed one scene with me. It called for the pilot to smash his head on the radio, putting it out of commission. I told them it'd probably put the pilot out of commission, too. If they didn't want to kill the pilot, they should find some other way to smash the radio. They agreed and changed the scene to have something loose in the cockpit fly up and smash it.

They said that for the studio shots they needed an airplane with two wings. Then they said, well, we really only need an airplane with one wing. They said, well, we really only need the fuselage and we can kind of fool around with putting the wings on in the studio. It's supposed to be wrecked on the side of a mountain. It has to be an Aeronca Champion.

"Do you have anything like that lined up?" I asked.

"No," they answered.

"Do you want me to get it for you?"

They smile and said, yes, would you please. I take their names and address of the studio. I thought, gee, this is an easy way to make some money. I'll just call up Nagel over at Torrance Airport and pick up an old fuselage or wrecked plane. I'll buy it and rent to the studio or just rent it for the time being.

I called Nagel. Well, he hadn't had an Aeronca Champion in 18 months. I began to worry, because if Nagel didn't have one I didn't know where I could get it. I called everyone I thought might help me. I was told of one here and there, only to find it sold. I discovered that, except for those completely demolished, the little planes were in backyards being rebuilt.

Oboy, this was going to be tougher than I thought.

I was at the Long Beach Airport that same week and saw Steve in his Aeronca Champion...and just about the year and model that the television people wanted. I thought of saying something to Steve but decided to wait; maybe we'd work something else out.

I went up with Steve on the flight check. He was a complete flop. He had about 250 hours logged and should have been trying for his commercial license and, instead, he couldn't even pass his private check. He was dangerous in the air, in fact, and almost clobbered us on landing.

He had the audacity to ask, "Well, did I pass?"

"No. One of the rules is that you're supposed to land properly and not crash on landing or you don't get your license."

He was insulted. Then I told him he was to get 20 hours more instruction and flight time and that I wanted to see his flight instructor. Steve said he wasn't going to use the same flight instructor; the man weighed about 200 pounds. This made the plane grossly overweight with both men in it.

"Okay, why don't you use one of my instructors and when he says you are ready, I'll go with you for another flight check."

The next day he was out with one of my instructors, shooting landings at Long Beach Airport. There was some cross-wind and Steve used left rudder on the left cross-wind. The instructor tried to override him on his own set of controls but he couldn't get enough right rudder and the plane went off the paved strip into the field. This is all right at Long Beach because there is lots of open grassland. But then Steve put on the brakes, which wasn't all right. It flipped over on its back.

Well...well. Here we had an airplane. I asked Steve if he wanted to sell the plane. He said, yes, without the radio he would sell it for $700.

I didn't know if this was a good price for a wrecked Aeronca, but I needed that Aeronca.

"Okay, Steve, I might be able to make a deal for it."

I phoned my man at the studio and told him I had found an Aeronca and thought it was wrecked exactly the way he wanted. He laughed. I told him that I could buy it and then rent it to them if he liked, so we talked price. I had large figures in mind that I had been given by people who dealt with the movie studios, but TV is a little different. We finally agreed on $700.

We settled on $700, not because that is what I could buy it for but that's what the studio workers said they could build the fuselage for. I told him that I'd make out on the deal by buying the plane, renting it to him, and then selling it afterwards.

This was all done on the telephone, nothing in writing. I've found that movie people are dependable and a deal is a deal.

The studio man was to leave the next morning for Mount Hood, Oregon, to do more scenes where part of the story had already been shot. I offered to meet him at the airport before he left with pictures of the airplane.

"No, that's not necessary. If you say it is the kind of plane we want, Claire, that is good enough for us."

I called Steve immediately to tell him I would buy the plane, that I wanted to meet him in Long Beach that night. He said, fine, he would meet me.

I took two checks with me instead of the one I'd normally have taken, because I had this little plan....

When I got there I made out the bill of sale, the registration papers, so forth, for him.

Then I suggested that it might be a good idea to put some money on account and learn to fly in a nose wheel airplane, in one of my Cessna 150s. A tail wheel airplane had him snowed; he didn't have the ability to learn to fly one properly and there wasn't any point in his trying to do so any longer.

He agreed. "Maybe I should put a couple of hundred dollars on account."

"Great. I brought along two checks. I'll make out one for $500 and one for $200, and you sign the $200 check back to me."

All of the $200 was soon eaten up. He needed lots of work and so the $500 was used up, too.

In the meantime, my mechanic in Santa Monica rented a trailer, went to Long Beach, took the wings off the Aeronca, and trucked the whole thing up to Santa Monica, unloading it in a corner of my hanger.

A number of people looked at it and made offers. I told them that it would be for sale in about 30 days.

The studio man came over in a trailer-truck and hauled the plane to the studio. The wings were put back on and the scenes shot. In ten days they brought the plane back to the hanger and the check for $700 arrived a few days later.

I sent the propeller to the local prop shop; it was refinished for $54.

Now I was ready to sell the plane. A friend wanted it for "sentimental reasons," saying he had once flown it, and offered me $400 for it. I turned him down, as I did others with similar offers. One said, "But I have money in hand!"

I said, "The airplane hasn't cost me anything yet; I've about broken even on it and I expect to sell it for what I paid."

One evening I was at the Orange County Chapter of The Ninety-Nines where I presented my Australian talk and films (I'll tell you about that in another chapter). Nancy Crews was there with one of her sons. She told me about the Taylorcraft they had in their garage and what fun it was rebuilding it. I told her about the Aeronca Champion.

Nancy looked at her son and drawled, "Boy, do you think we dare buy that and put it in our garage? If we do, we had better not tell Pappy about it. He's sure to be unhappy, having two airplanes in our garage!"

It seemed "Pappy" didn't fly. Nancy was an ex-WASP; she did some instructing and towed sailplanes at a desert gliderport.

In a few days, she, her husband, and a friend who was helping her rebuild the Taylorcraft, came to look at the Aeronca. They bought it for $800. That covered my purchase of the plane, fixing the prop, the trailer charges, and the mechanic's pay. The TV rental for $700 was my profit.

While all these events were taking place, Steve — the former Aeronca Champion owner — had finally learned to fly. And I finally passed him on his flight check. The day I wrote out his private ticket he bought one of my used airplanes, one with a tricycle gear.

Chapter 10

What is it like to be a flight instructor? Let me take you through some incidents from the thousands that have occurred to me or to my fellow instructors.

The Cessna 150 engine drone softens as my student pulls the throttle back. My young student is in the pilot's seat while I sit on the co-pilot's side, the right-hand side of this two-place trainer. His hands tremble slightly, his forehead dampens as he concentrates during his hour of instruction.

My flight school's practice area is behind us. We keep the almost solid line of houses along the Malibu shore on our left. I voice-guide the young man's hand as the plane descends from three thousand feet to the twelve hundred-foot turning altitude over the Marina del Rey's ocean entrance.

A lazily undulating white foam line divides the gray Pacific Ocean from the gray shore sand. A light brown haze fills the Los Angeles basin below us. I glance far inland where mountain peaks stand tall and deep brown in clear relief; their lower sides murky. The plane descends with reduced power into the smog.

"You are doing just fine. Keep it like that—bring your right wing up just a hair—easy now—throttle forward as you bring the nose up to level—that's fine, hold it there."

I am constantly looking through all the windows. "Always keep an eye out for traffic. All clear?"

My student looks in all directions and nods. "Yes." "Okay, we're almost to the marina now. Pick up the microphone."

I watch the student wrap his hand tightly around the mike base, pull it from its hook on the instrument panel and bring it to his sweating face. As we near the marina—the radio reporting point for the Santa Monica Airport downwind leg—I tell him, "Now I want you to say, 'Santa Monica Tower, this is Nine-One-Uniform; we are over Marina del Rey.' Then you request landing instructions."

Muscles stand out in his shirt sleeves as he squeezes the mike button just below the speaker. "Santa Monica Tower, this is Nine-One-Uniform. I'm over the marina with Claire. I want to land."

I hear the Tower microphone clatter as though it had been dropped. The voice of the Tower operator comes through the

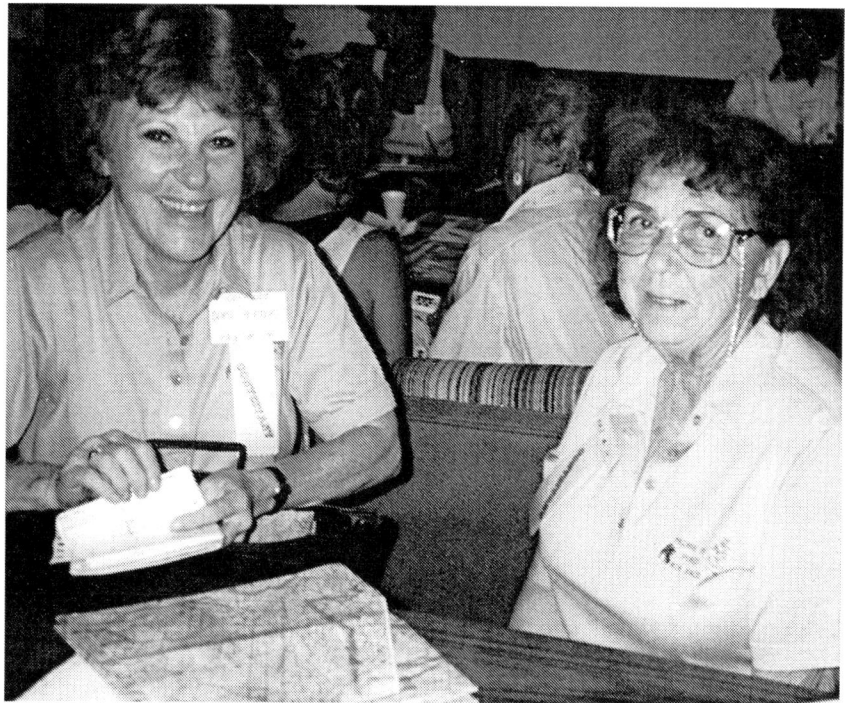

Emma McGuire (right) was one of the finest flight instructors Claire ever knew; she worked for Claire for many years. At left is Doris Minter, a pilot who races often, and often volunteers to help the Palms to Pines Air Race.

overhead speaker with a low and paternal inflection, choking on laughter: "Come on home, Claire!"

I chuckle and direct my reddening student into the 45-degree angle on which we are to approach the airport, then we turn to stay parallel with the airport runway on the downwind leg. I carefully instruct him on the landing pattern procedures, have him call the Tower to report our position. The Tower operator responds, directing us to land third, behind a twin Beech.

We turn now onto the base leg, then onto final, facing into the wind. I see the student glance at me, obviously wishing I would take over. This will change in another three or four hours of instruction; after that he, if typical of male students, won't be able to conceal his desire to solo.

"Just keep your hands and feet light on the controls. Keep your toes off the brakes. You're doing fine!"

We glide over the high embankment rising above the traffic intersection at the end of the runway; the plane lowers slowly and

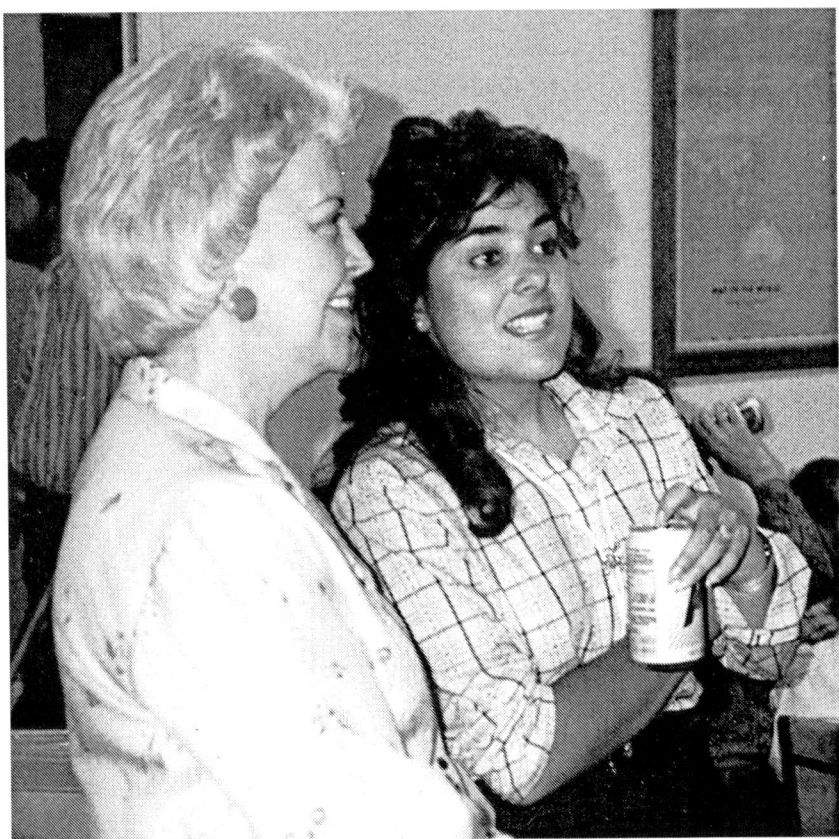

Margaret Callaway (left) taught her two daughters to fly, as well as many others. Loves to air race.

touches gently on its main landing gear. The nosewheel touches a few seconds later. I grin and nod approval; my hand relaxes. We move off the runway.

The student picks up the mike — good boy! — identifies our plane to the Tower, and reports the position as he toes the rudder pedals, turning the plane onto the taxiway. He taxis slowly, his head moves back and forth as he watches out for other ground traffic in the busy lanes. He turns the plane into the tie-down space opposite the Claire Walters Flight Academy sign atop my offices, pulls back the mixture control, and turns off switches.

We check the instruments on the panel; on my clipboard I write down the time and some remarks about the flight. We get out, walk around the plane, then each of us picks up a tie-down rope attached to an iron ring in the asphalted ground and run the end through the small ring on each wing, tie the rope snug at the iron ring. This will keep the plane from rolling in the breezes.

As we walk to the flight office we discuss the lesson. At the counter in the spacious lounge, I help him fill out his flight logbook. I sign it, suggest he study a particular section in his flight manual before the next lesson. He agrees, walks over to the orange-colored couch, sits down and puts his notebook on the magazine-filled coffee table in front of him.

I look around the room. Three small offices, one for my bookkeeper, one for me, and one for any flight instructor who needs to talk privately with his student as they plan the lessons and cross-country flights, open off the customers' lounge. Large framed aviation photos gathered over the years are hung in a long row on the wall space above the doors.

Arranged on the hallway-side wall opposite are art objects that I have collected in my foreign travels, many from Australia in 1966 at the end of my flight across the Pacific Ocean from Oakland, California. I flew a twin-engine Piper Aztec solo, in a flight of three small planes, to deliver it to Avis Rent-a-Plane in the Land Down Under.

One of my flight instructors is hunched over some papers at the long counter. I ask, "Is he off all right on his cross-country?"

The instructor looks uneasy. "Yes, and I think everything will go okay. We've made a number of good local flights and we've made the identical triangle he's following today. He plotted the course again this morning and did just fine."

"What's the matter then?"

He just shakes his head and spreads his hands helplessly.

I turn away and study the day's flight schedule placed near the small telephone push-button system. It is booked nearly solid with private and ROTC students; my five instructors are busy today.

A student walks briskly through the hallway door, greets me and the flight instructor. The two men go into the side office to discuss the flying lesson coming up.

I glance over the items in the glass-topped display cabinet at which I'm standing. It is filled with logbooks, flight manuals, navigation aids, and various items of aviation jewelry. On the walnut-paneled wall behind the cabinet are stepped shelves containing my flying, softball, and pistol-shooting trophies, as well as a few of the prettier vases from my collection.

One of the school's two women flight instructors comes through the open door at the far end of the lounge. Her male student follows. We wave at each other; they drop their charts and notebooks on the large worktable near them and walk into the small catch-all room where a coffee urn is perking.

I go into my office. Jean Grooms, bookkeeper and general all-around assistant, comes in with some papers for me to look over, explains them, and goes back into her own office next door. She's been with me for a long time. Dependable, responsible.

I fiddle with one of Jean's invoices but I think of the young man on his solo flight cross-country. I look at the phone, dreading the moment it is going to ring.

Conscientious flight operators hear a strange, foreboding ring to the phone when a student is off on a cross-country flight...one knows he is on the other end of the line, or someone else is calling...and the news won't be good.

Most of the time everything goes off without a hitch. If the instructor has the slightest doubt about the student's ability to succeed solo, he doesn't send him out alone. But no matter how much one teaches and tests, no instructor can be absolutely sure how a particular student will react in a strange situation. And this student, well....

This student is a young man from the slums of New York. Barebones schooling; poor family background. His high, cracking voice, laced with the "ese" of his origins, makes him sound like the picture looks that everyone has seen hanging on walls — the double-featured boy who says, "Six months ago I couldn't spell 'instructor', and now I is one."

He'd been passed from one instructor to another as each gave up trying to teach him to fly, to understand and trust the instruments, to work the radio. He just couldn't get the hang of it. Finally I took him over.

Through sheer patient perseverance, over and over and over and over, I was able to get him soloed.

I could see that the young man had many good qualities and good potential, but only a lot of hard work was going to bring them out. He had never ridden a bicycle or motorcycle, driven a car, nor worked with tools.

Except for hitchhiking across the country from New York to Los Angeles, he had seen nothing of the country, knew geography only generally. He could not seem to relate the lines and marks on the chart with the way the terrain looked over which he was flying during visual flight work. But he refused to stop trying.

He wanted to be a pilot badly — but he was, after many hours, still just a bad pilot. Until the day when the dawn started to break in the boy's head. All the bits and pieces started to fit together, or so

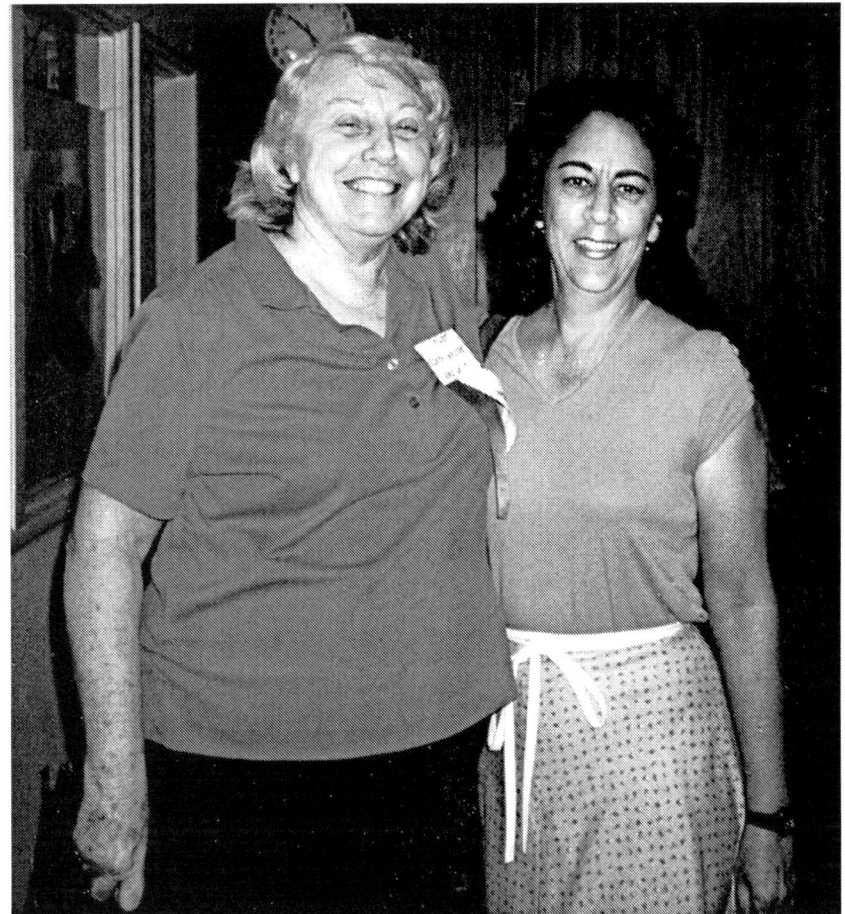

Elizabeth Dinan (right), flight instructor, loves to air race. Shown with Claire.

it seemed, and everyone cheered when he soloed. At that point I turned him over to another instructor. He was to prepare the student for his cross-country flight work.

Now he is, finally, on his solo cross-country flight. He is to go to two different airports a reasonable distance away, check in at each, and return home. The two airports chosen were Bakersfield and Fresno, both north of Santa Monica with a pleasant flight over fairly low mountains and through the long green San Joaquin Valley. North of Santa Monica Airport.

I finish a business call and hang up the receiver when the phone rings again. Again my heart sinks.

The high, cracked voice said, "Claire, have you ever heard of George Air Force Base?"

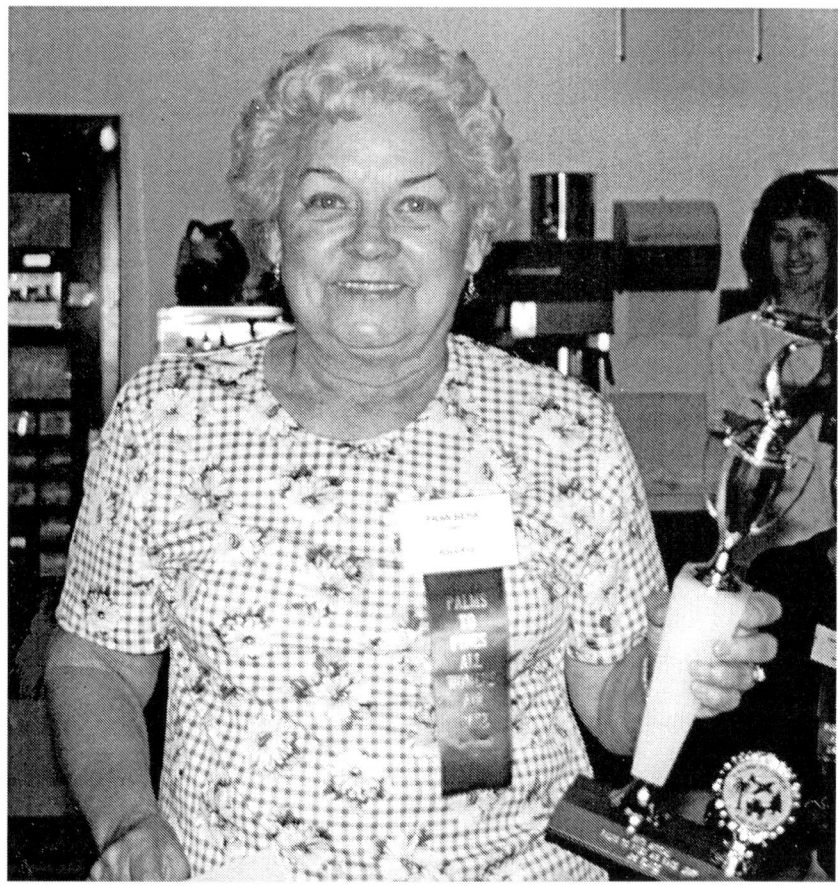

Frances "Fran" Bera, 50-year pilot and long-time instructor, was seven-time winner of the Powder Puff Derby. Collects winning trophy in Claire's Palms to Pines Air Race.

There are 4500 women flight instructors today. There were less than 100 when Claire decided she wanted to be one of them.

"George!" I gasp. "What are you doing at George?" George AFB is 90 degrees to the right of his course, out in the Mohave Desert!

"Well, I couldn't figure out the Omni and I couldn't figure out the compass. The compass said one thing and the Omni said another, so I just split the difference and I ended up at George Air Force Base." He paused for breath.

"Everything's okay, it's just that they don't have 87 octane gas here; they have 130."

"Take it, take it. Just take it and fill the tank!" I laugh in relief. Migod! Suppose he'd split the difference in the other direction —

he'd have been over the Pacific Ocean! Luckily he wasn't color-blind or he just might have!

"Do you know the way back home?"

"Ya, ya, Claire, no problem, Claire, I come home."

As soon as I hang up I regret not grounding him right there. The clock seemed to stop and the stomach butterflies began again. The flight instructor comes back with his latest student and leans against the doorjamb of my office, his eyebrows raised in inquiry. It is my turn to gesture helplessly. "I'd better give up — call him back — tell him to stay there. A couple of you can fly out there and bring him back."

The base officer-in-charge came to the phone. He told me that the student was in the air and headed in the right direction for home.

"I talked with him for a long time and made sure he understood me. I think he'll be all right."

I thank him and hang up. Still I sit, shuffling invoices, wondering "what next?" An hour crawls by. He is due. He is overdue. Where is he?

The phone rings; I flinch.

The high, cracked voice said, "Claire, have you ever heard of Norton Air Force Base?"

"For cris'sake, what are you doing there?"

"When I came over the mountains I saw all that haze and it looked pretty solid to me. I looked down and there was this big airfield, so I landed. What'll I do now?"

"You stay right there. I'm sending a couple of guys up to get you!"

I send two instructors up there after him in another plane, because, after all, I would still have had March, El Toro, and Los Alamitos military bases left to deal with.

We start out teaching our students to navigate by visual landmarks in conjunction with the compass. Freeways, water tanks, railroads, mountain peaks and valleys, shapes of cities or some landmark within the city, such as a racetrack.

There was always smoke and haze in the Los Angeles Basin. It was better in Santa Monica because the sea breeze would push it back to the San Gabriel Valley.

At the time I operated my flight school, we had installed in our planes the VOR (VHF Omnirange). The VHF omnirange operates within the 112-118 megacycle band. In this band it is relatively free from atmospheric and precipitation static and interference from other radio stations. Furthermore, it is not limited to four courses as is the A-N range, but provides definite guidance on any course, to or from a station, that the pilot may select.

That is why it is called the omni (directional) range. At minimum instrument altitudes the VOR gives reliable indications up to about 50 miles, depending on enroute terrain.

It did not take long for the average student to learn to use the Omni, and as soon as he got the instructor out of the cockpit, to switch it on and forget visual navigation. This wasn't good, of course; Omni stations sometimes go out, or weather closes that field and one must find another field, only there's no facility....

Too many instructors fail to teach chart navigation properly and to use the VOR as a back-up.

One young man went off on his first solo cross-country, only to return 30 minutes later. He came into the office white as a sheet and confessed that he'd switched on the Omni as soon as he left Santa Monica Airport and aimed it for Phoenix, Arizona. He looked down 15 minutes later to find himself flying over the Pacific Ocean, having read it 180 degrees wrong.

First-time racers Shelly Funk and Summer Fields of San Luis Obispo, California, flying in my Palms to Pines Air Race, tells us that on take off from Santa Monica, they thought they'd be ahead of the competition by using the LORAN (Long Range Navigating system) to make their navigation easier through the unfamiliar Los Angeles Basin. They were not yet over the end of Santa Monica Airport when the LORAN told them they were 99 miles out.

Currently, the GPS (Global Positioning System) is all the rage among pilots. But it has the same built-in problem: pilots tend to believe in it blindly. But the gadget they can hold in their hands which tells them exactly where they are on the planet, can be faulty in itself, the batteries in it can go dead, or the satellite itself, bouncing the beams, can be hurt by some space happening. The paper map on your lap, checking it by an occasional look out of your window, should never be out of fashion. Besides, you may need a landing place suddenly.

Years ago I was instructing a student in a flying club's airplane. It was one I had spent many hours in with students. We were on a dual cross-country to Bakersfield, where we made a landing for fuel. I noticed that the engine was heating up and I mentioned this to the student.

"But, Claire, the engine has just had its 100-hour [engine inspection and work] and it's been checked out real good. It should be all right."

"Well, okay, we'll see. It is a hot day."

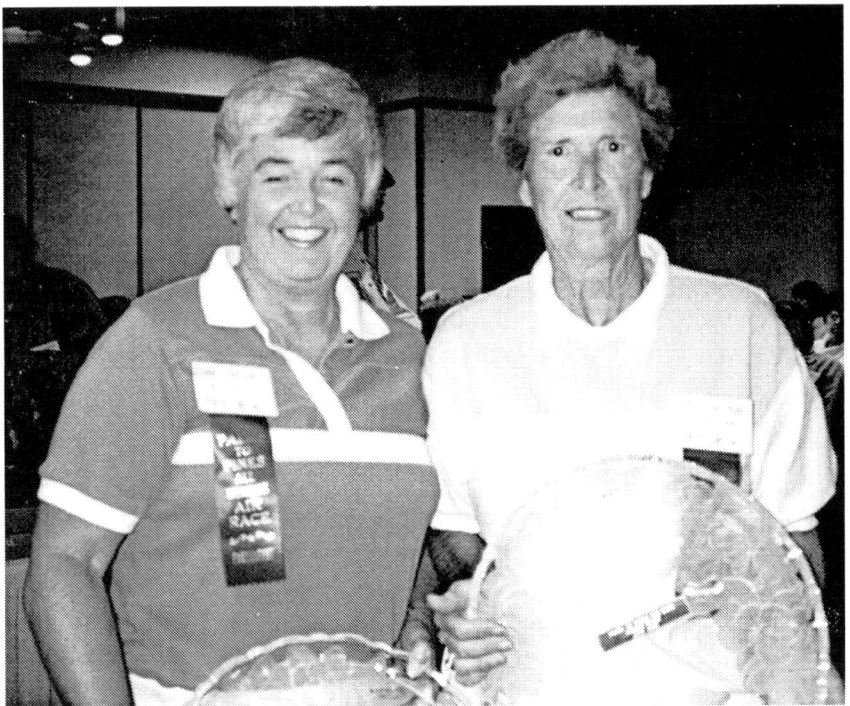

Donna Taylor (left) and Jane LaMar are holding trophies from their "upteenth" Palms to Pines Air Race. They are nearly always a team. Jane is a flight instructor and racer.

We left the airport area and, at 6000 feet altitude, headed towards Santa Barbara. The coastal mountains were visible and in a few more minutes we'd be right over them.

A terrible banging then as the engine cracked.

The student said, "Oh migosh, is there an airport around?"

"Yes, right below us, 5000 feet down."

We spiraled right down over the field. I called the Taft Unicom to say we were making an emergency landing. "Okay, all clear to land," they replied.

Well...it was Sunday afternoon in Taft, on a hot, hot day. Nothing moved in Taft. The trains didn't run, the bus didn't run; there was no way on the ground out of Taft. We phoned our home airport to send another airplane for us.

Just had a call from the Santa Monica Tower. Seems one of my students from India was coming in from a solo flight and asked for landing instructions. The Tower told him to fly around the pattern again, but the young man's English "ear" isn't good enough: he flies around the Tower! I am to speak to the Indian as the Tower occupants do not appreciate this. I agree and have to chuckle as I sigh. Many of my foreign students come to me via the courtesy of the U.S. Department of State. I must be gentle.

The Tower I speak of is exactly that: a lofty glassed-in eagle's nest sitting near the runway, filled with radio equipment, radar scopes, and men and women who guide flying traffic off from and onto, and near, the airport.

While some of us in the office were watching a DC-3 making a couple of instrument passes at the field — one of my instructors was checking out someone on his instrument procedures — I told them about the time I flew in a DC-3 with Robert Macafee.

"Mac" had been the personal pilot for John Kennedy before he became president. He also had flown Johnson before he became president, and piloted two men who later became Catholic popes.

"Mac" came out here to work for Aircraft Associates at Long Beach. I first met him when he walked into my office one day and asked to be checked out in a Cessna 310, a twin-engine. We learned to have a great deal of respect for each other.

Aircraft Associates had a Cessna 310 on lease, and "Mac" wanted to get checked out in it. I tried to find an instructor for him, but he wasn't available when they were.

He came in one day and said, "Claire, I'm only going to fly with you. You're going to have to check me out. Nobody else."

It thrilled me that he wanted me to do it. It took just a little time for him to get used to the feel of the Cessna 310. He was an excellent pilot.

He took me along on a couple of trips to Texas in a DC-3 and I got to log some DC-3 time as co-pilot for "Mac."

Chapter 11

Our primary focus was on flight instruction from fledgling to multiengine airline transport rating. We instructed in single and multi-engines and for instrument ratings in both, and taught wanna-be flight instructors. We filled in with charter flights. Planes flying earn their keep.

A man who had flown with us before phoned me from Ontario (California) to charter a Comanche to fly from Ontario to Las Vegas. He said we'd have dinner there, spend the night, and return to Ontario the next morning. He had some business deal where he'd gotten two business people together and he was to collect his commission.

My business partner for a time, Betty Faux, and I flew to Ontario to pick up him and his girl friend. It was obvious he had been drinking. We collected the charter fee and prepared to leave. He insisted on the co-pilot seat. Against our better judgment we let him. I sat in the pilot's seat (left side) while my partner sat in back with his girl friend. He wanted to take it off so I let him but I was forced to override the controls; he was sloppy.

We reached Las Vegas and registered at a hotel. We were all to meet a little later for dinner.

Our client went into the casino to cash a check in order to gamble as he wasn't carrying much cash. The casino refused to cash it, acting as though they didn't know him. When he phoned the business people he'd brought together, he couldn't find them.

This often happens to business "finders." If they don't watch their clients closely they'll be shoved out of the deal.

He was mad and changed his mind about taking us to dinner, saying we were returning to Ontario that night.

We went out to the field. I told my partner to hurry and get in the pilot's seat before the man took it. He was mad because I made him get in the back. I took the co-pilot seat. We found Ontario Airport hazy. A great number of airliners were there — a signal to us that Los Angeles International Airport was blanketed with heavy fog.

We reached Santa Monica and found it covered with dense fog, but so narrow a band that we could see tree tops above it. In inland valleys this is called "tule fog". We couldn't land and turned towards Van Nuys where it was clear.

We taxied up to the tie-down space. There were a great many other airplanes stuck there for the night.

An old man, about 75 years old, watched us get out of the plane. It was about one a.m.

The old man said, "Would you ladies like to have a drink?" Betty Faux said, "Sorry, we have to go to work this morning."

He was dressed in a pair of white ducks, a blue shirt and jacket, and had a small beret on his head. He cocked his head and said, "I'm waiting for my pilot and plane to arrive so I can get to Palm Springs." He hesitated a moment, then said, "Could this airplane fly to Palm Springs?"

"Sure," I said.

"Would you fly me to Palm Springs?"

"Yes." "Well, if my pilot doesn't show up pretty soon I'll have you fly me to Palm Springs."

By this time my partner was giving me good long, hard looks. She was plenty irritated with me. Well, we were both tired, but could we turn down a quick buck?

I said, "Maybe we'd better have a cup of coffee, then."

At this point another man appeared. The old man turned to him, handed him a small card, and said, "Go call this number for me and see if the pilot is coming."

When the other left to phone, the little old man turned to me and said, "Why don't I let you keep the money for the charter now while we see about the other pilot. How much will it be?"

I said, "Hm, hm, now, let me see...", doing some rapid calculations. I was just about to say, "$60," when he said, "Would a $100 be all right?"

I've been glad sometimes that I pause before answering about costs. "Yes, that will be fine."

He handed me the money. The other man came back and said that he had reached the pilot at home and told him not to come.

The old man said, "Okay." Then he turned to us women and said, "Now, girls, I'm going to stay at Jimmy van Heusen's place tonight in Palm Springs. Why don't you just stay there and fly back in the morning? You're welcome to stay."

I declined. "Thanks, but we have to work in the morning."

As we walked out to the plane, the other man tagged along. I said, "Why don't you come along? We're coming right back."

The old man said, "Will it cost extra?"

"No."

So we climbed in. I got in the back with the other man while Betty sat in the left-hand pilot's seat, with the old man on her right.

Every once in awhile the old man wanted to take the controls. He'd jerk the plane around for a bit then hand it back, saying "Ah, there! Now I have you back on course!"

During the flight out I found out from the man beside me that the little old man was head of the music department at MGM (Metro Goldwyn Mayer) movie studio. I asked him who he was? A friend? Did he work for him?"

"Oh, no. I'm a cab driver; I own my own cab. I bring him out to the field from Malibu whenever he wants to fly to Palm Springs."

We got to Palm Springs and landed. The little old man climbed out. He paused and looked back in. "Will there be any more money due?"

We said, no, that was all. We all said goodnight and left, he for an airport taxi, us for Van Nuys.

We got back to Van Nuys and found that Santa Monica was still socked in. It was pretty late, about 3:30 a.m. We were standing by the plane wondering how to get to Santa Monica.

The man spoke up. "Well, now, you have forgotten I own my own cab. You gave me a free ride to Palm Springs; I'll give you a free ride to Santa Monica."

So that's how we made it back to Santa Monica.

That night we had made about $265 on charter flights — one to Las Vegas and one to Palm Springs — with a free cab ride to Santa Monica. One Sunday morning we got a call from a man who said he wanted to go to Las Vegas that night. This was quite unusual. Sunday is not the day most people fly to Las Vegas. They go Friday night or Saturday night.

It was a man who had taken a couple of lessons with us. He wanted to leave at eight that night. My partner and I went out to the field and checked over the Cessna 172. He didn't show up. Finally I called. He sounded drowsy and perhaps drunk.

He said he'd be out to the field at ten. He came and sat in the left seat while I flew from the right seat. My partner sat in the back. The two of us always took charter flights together. We'd found that it was safer, especially to Las Vegas, since that usually meant handling drunks.

We took off and headed toward El Cajon Pass. As we neared the pass he turned to us and said, "When we get out in the desert, I'll tell you where we're really going."

By this time it was 11:00 p.m.

I said, "No. You are either going to Las Vegas or you are going back to Santa Monica."

He repeated, "When we get out into the desert I'll tell you where we're really going to go."

"I think you'd better tell me now what you have in mind." "Wait until we get into the desert."

My partner smoked her cigarette more furiously, as though she intended to stick it in his ear.

Finally he dragged out a map. "We're going to B___, Nevada."

"No, we are not going to B___, Nevada." I said. "We won't get there until midnight and I've never been there. I don't know what it is like and I'm not going in there."

"It's got lights on both ends of the runway. There's a house on a ranch strip and that's where I want to go."

"Absolutely not. Either we go to Las Vegas or back to Santa Monica. Why do you want to go to B___? Do you have a business deal there?"

"Yes."

"You knew about it this morning? Why didn't we go earlier?"

He didn't say anything.

We flew on into Las Vegas and landed. "All right," I said. "We'll take you into B___ tomorrow morning."

The next morning he didn't meet us in the lobby as he said he would. We had the room clerk call him. He came down, didn't say anything to us, and headed for the bar. We were angry.

I went over to him at the bar. "If you want to come with us back to Santa Monica, you can, but you can't drink."

"I'm going to have this drink."

"Then you'll have to stay here. We're going back to Santa Monica."

This was what I told all my instructors. They were to collect the money for the charter trip in advance. If it doesn't work out at the other end they are to come home.

I told him we were going to take off in 30 minutes. If he showed up he could ride home with us. He didn't show.

I was telling the story to some people later that day. One of the fellows said he was going to check it out and see just exactly what was going on in B___, Nevada. He called me later and said there was a house of prostitution at B___, and business men flew their customers there for a little entertainment. That the fellow in my plane planned to buy the house, and wanted to fly into B___ without notice to find out how much action there was.

The man had told me nothing about that. I'd no idea that the place existed. I had imagined all kinds of things — smuggling, chasing down a runaway wife, other kinds of trouble.

I'm certainly glad we didn't take him there. We might have run into someone we knew. It would have been difficult and embarrassing to explain what we were all doing there.

Las Vegas charter flights can mean trouble. Occasionally I had a flight instructor who didn't follow my instructions and come right back: I'd point out the facts — we need him and the plane back at Santa Monica earning their keep.

He needn't stay more than few hours waiting out of concern for a passenger who has not shown up when he said he would. It doesn't cost that much for an airline ticket.

Some years ago I flew a young couple (he was a medical doctor-psychiatrist) out to Inyokern to see about a job. They were to meet with the clinic officials near China Lake.

The young wife, as she stood near the Cessna 172, expressed some doubts about this flying jazz. She started to climb in the back and so did he. I said, "No, I need one of you to sit up here with me. The airplane will be better balanced."

His wife said, "I don't want to sit alone."

"Then you get up here with me."

So she did.

He said he had his own boat and he began to do a little navigating. He was doing well at it.

She got interested in the map so I showed her how to read it. She began to find locations and soon began to tell him about certain symbols. She located our position and was elated.

She became enthusiastic and talked about learning to fly. She had began the flight as a timid young lady and ended it with the desire to learn to fly. She showed a great deal of promise, at least, in the navigation department.

It was a beautiful day, clear, crisp and with little turbulent air.

We landed at Inyokern behind another plane. As we taxied up to the line and stopped, some men came over to meet us.

One said, "We saw two planes come in together and we didn't know which one you'd be in until we saw you taxi. We thought, yours was the plane whose pilot seemed to know what he was doing so we came over here."

This is the nice kind of charter trip which we didn't mind doing; nice people to be with on a beautiful day. The bad part of charter flying is the waiting for them at the other end of the flight...usually waiting for hours. If someone asked us to take him frequently, we'd ask him to learn to fly — we'd teach him.

Some years ago while we were up instructing, a fellow came into our office and rented one of our airplanes. The boy behind the desk was doing it properly: he had the man renting the plane fill out a checkout card, arrange to be checked out in our airplane, and asked him to make a substantial payment.

Unfortunately, another employee intervened. Because this man was brought to the airport by a friend of hers, she thought that he was all right and insisted on my office letting him have an airplane. Without checking out with an instructor, he flew away in our Cessna 172.

He had signed some name and left $25 as a deposit on a flight supposedly to Palm Springs and back on a real estate deal..."a million dollar deal."

I became suspicious and phoned various flight schools and found that he had rented an airplane at one school, and had stolen one from someone else. He had quite a reputation, all bad. He was an ex-convict who had done a lot of bad things, including stealing airplanes and wrecking them.

I phoned about the country. I found our plane at Las Vegas. The manager there blocked the airplane with his after putting the airplane in the hanger. He blocked the front of that with rental cars. We took an airliner to Las Vegas, got there early in the evening, and checked the plane; it was all right. Then, we looked for our man but never found him.

That night I was sitting in the airport waiting room hoping our man would show up when an elderly man wandered into George Crockett's flight office. He had tried to buy a ticket on an airliner going to Santa Barbara, but he was too drunk. He had come into Crockett's to see if he could get a charter flight to Santa Barbara.

The manager sold him a ticket to Santa Barbara and said they would take him in the morning; they wanted him to sober up a bit. They rented him a room there at the motel.

I suddenly knew how I was going to make our gas money home. When the old man had gone to his room, I told the manager, "You know, we have to get back to Los Angeles tomorrow morning anyway. If you have a charter flight that you don't particularly want to make, we could take it and just split the dough with you. You can make money without turning a wheel."

He said, "How about you taking the old man to Santa Barbara?"

"It's a deal."

He gave us half of the ticket money and in that way he made a neat profit without ever sending a man out, and we agreed to take the old man at 6:00 the next morning.

We all met in the office, poured the old man into the Cessna 172 and off we flew to Santa Barbara.

He tried to converse with us, but he could barely talk. He had been drinking for so many years he was in a bad way.

We smiled at him and were friendly to him. And we'd look back once in awhile — he was sitting in the back seat of course — and try to talk to him; and he was very nice, no problem.

After a little while he wrote something on a card and handed it to me. It said, "You have such a sweet smile."

I gave him another smile and he handed me a $100 bill.

We got over Santa Barbara and he asked us to fly him over his house, that his seven poodles would want to know that he was home.

He directed us over the largest house in the big Santa Barbara wealthy area between the town and the airport, and we circled the house.

We landed at the airport and called him a cab. Before he left, he told me that as soon as he sobered up he was going to call me; I was his kind of girl. I guess he never sobered up.

John Gavin chartered my plane to go to a grape workers unionizing meeting in Delano, California. The workers wanted to determine whether to join a union, and if so, which union. Two unions were vying for the workers.

Gavin, at the time, was First Vice-President of SAG (Screen Actors Guild); later he was president for two terms: November 1971 through November 1973.

Gavin was asked to speak on behalf of joining a union. He speaks Spanish and Portuguese fluently, which stood him well when for five years, 1981-86, he was Ambassador to Mexico during the Reagan presidential years.

We arrived at the airport. Gavin insisted I come with him to the meeting, and when we got there he saw to it that I was seated next to him.

The leader of the meeting introduced him and told him that an interpreter would translate his speech. Gavin firmly said no, he'd do that himself. He spoke first in English and repeated it in Spanish. He was a big hit with the audience. He told the workers he didn't care which union they joined, just join one.

John was always a perfect gentleman. I always had a great deal of respect for him.

Other movie stars that I met in the years following were always gentlemen, too. Gene Hackett rented planes from me until he bought his own. Robert Hays received his private license from me. I taught Anthony Zerbe's 13-year-old son Jared to fly and soloed him when he turned 16. I gave him his private license when he turned 17.

We tried to keep sight of the idea that flying is a lot of fun, too, as well as a goal-oriented activity.

A major part of that program were Sunday morning breakfast flights or, sometimes, weekend flights to a special somewhere.

Over the years we had planned and carried out many of these flights and they had been reasonably successful, with many of our students as well as others participating.

The Buffalo Barbecue Fly-In was our most famous and most fun overnight flight. I decided that a western-style outing to Barstow, California, a small desert town on the road to Las Vegas, Nevada, might be the ticket. We'd sleep in sleeping bags under the wings of our planes and cook over campfires. We had done this on a similar flight to Giant Rock near Twentynine Palms.

I phoned a fellow pilot, Joyce Failing, who lived there. She was an attractive blonde who, during World War II, was a radio operator for the CAA on Silver Lake Airport, near Baker.

Joyce was a member of The Ninety-Nines and had served on the Board of Directors for the Powder Puff Derby.

She thought about my plan for a moment, then said, "Okay, we have motels and a nice airport here. Will you let others come?"

"Of course. People who own their own airplanes, or can rent one, those from other flying schools, or from anywhere, are welcome. We just want everyone to have fun. And, we get our planes rented out."

Joyce said, "Okay, let me think about it and work on it for a bit."

Every few days, Joyce phoned with a new idea. This was Joyce, and why I had called her. She was always a bundle of energy, a hard worker with lots of good ideas.

"Do you mind if people from other states come?"

"No, of course not."

"Is it okay if I go to the chamber of commerce and service clubs to see if I can get some help on a free weekend trip?"

"Fine."

Joyce was president of the Barstow Desert Area Chamber of Commerce at the time — the first woman president of a CofC. She served from 1960 to 1968. She had just returned from the national CofC convention in Las Vegas where General Motors had sent 100

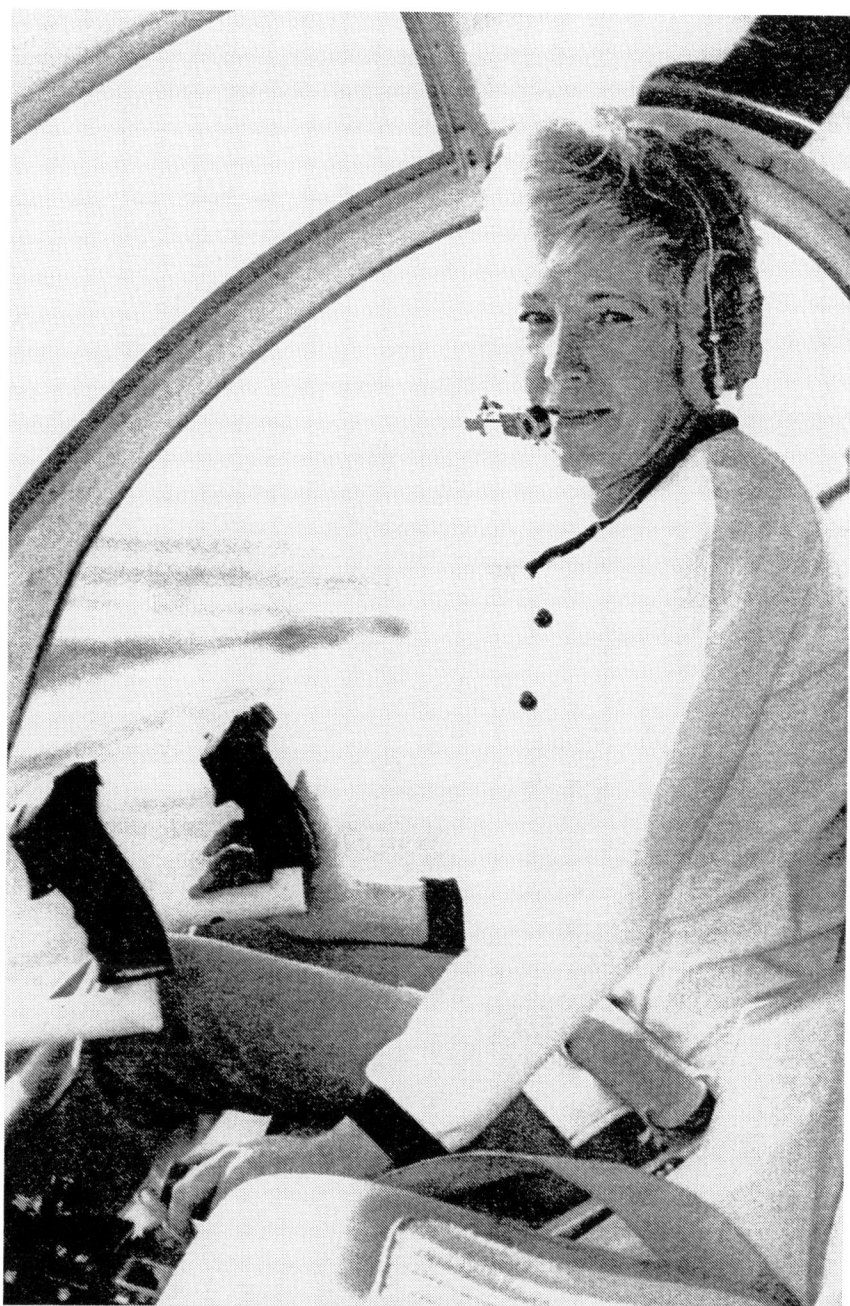

Joyce Failing — commercial, instrument, helicopter pilot with several thousand hours, was a mover and shaker of events in this century of flight. She was the first woman president of a chamber of commerce in the U.S., and the powerhouse behind the famous Buffalo Barbecue Fly-In.

Photo courtesy of Joyce Failing

white Buicks for convention use. They were still in Las Vegas. Joyce talked the company into letting us use them.

Joyce got members of the service clubs to go to Las Vegas and drive back the 100 white Buick convertibles.

She arranged for buffalo meat to be flown in from either Wyoming or Montana — I'm not sure which — to be barbecued at Calico, an old ghost town brought back to life for tourists by Walter Knott of Knott's Berry Farm in Buena Park, California.

She arranged for a dance in Calico, for a cocktail party, and for a stage show.

Joyce mailed information sheets to flying schools all over the West. As a result, 105 airplanes flew in from California, Arizona, and New Mexico. A lot of folks drove in.

On our arrival, we saw the long line of white convertibles at the field. A white convertible with a "Follow Me" sign on its back guided the taxiing aircraft to tie-downs. Boy Scouts helped push some aircraft into hanger sheds.

As we registered for the event, we were handed keys to one of the convertibles to drive to the motel where we were to stay.

We all drove out to Calico for the buffalo meat barbecue. Afterwards, we attended the special stage show for us. We returned to Barstow where we danced the evening away.

The next morning, Sunday, an air show was put on for us. Some new and some ancient aircraft did aerobatics.

I think that was one of the most amazing and certainly the most fun of all the fly-ins we have had. It was all Joyce Failing's doing.

Another fun weekend was a flight to Clear Lake, above San Francisco.

We had two seaplanes flown in from Sausalito and everyone had a couple of hours in one. We have flown to Sedona, in Oak Creek Canyon, Arizona. That area is absolutely beautiful. We'd have dinner together, go hiking, dancing, take photographs. They were always wonderful weekends.

Other flights were breakfast flights to Apple Valley, Santa Ynez, Catalina Island, Santa Barbara, Bakersfield, and San Diego. There were many quick flights up the coast to Santa Paula Airport for pie and coffee.

Many times my instructors — those who planned ahead — could use these weekend flights with our leased aircraft. Students could complete their solo and dual cross-countries while enjoying camaraderie with other pilots.

Usually these required cross-country flights were planned triangles — home base to two other airports and home again. The instructor and student could take off early and reach one of the points first, then go on to the breakfast fly-in point, then home, or go to the breakfast first, then on to the next airport.

This always makes it a little more interesting for the student. This went well with the family man. He could take his dual cross-country with his instructor in a four-place airplane, taking his wife and one child along. They'd experience the fun his flying could mean for them after he received his private license. And, just maybe his wife might want to learn to fly, too.... Hey, a flight school owner had better be thinking all the time.

Here is the chance for the instructor to meet the wife, perhaps talk her into coming out to take a demonstration flight alone with the instructor, where she can be talked into taking some lessons so she can take over if her husband becomes ill in flight. If flying scares her, she can get over it and go on other interesting weekends with her husband. Family flying can be a great deal of fun for all.

Sometimes it was the husband who would arrange for me to meet his wife, perhaps at a social gathering. His wife was protesting the amount of time and money he was spending on flying, say, but wouldn't fly with him. I'd talk her into a demonstration ride, then into a few lessons.... Months later her husband would protest that now that his wife had her license, she was taking the plane all the time and entering women's air races!

Chapter 12

It is a warm summer day in the late 1960s. I unlocked my suite of offices on Santa Monica Airport and helped my mechanics and instructors get the day's work started. I heard the noise level rise as the public room filled with pilots and visitors.

Instructors and students spread out charts at side tables and prepare to study them for that day's cross-country flights. Others stand at the glass counter arranging for planes or purchasing items from the cabinet below.

My own task today, besides the never-ending paperwork, is a low altitude photo run near the Lancaster Freeway, south of the Gorman area. The photographer needs aerial photos of some real estate properties. He hurries in carrying his camera equipment.

We get the Cessna 182 Skylane fueled and pre-flighted. Meanwhile, a friend, Gene Mauch, comes into the office. He is a professional with all the ratings, steady type, solid citizen, an airline pilot. I ask if he'd like to make the run.

"Sure. Sounds like fun."

I am at my desk finishing the paperwork when the phone rings. It is Gene. He is down on the Bakersfield Freeway. He is sorry, says he must have done something wrong, but doesn't know what. I reply that whatever it was, I'm sure it wasn't his fault.

Gene says that the engine had simply gone into idle, hadn't quit. He was about five hundred feet above the ground, with hills around. He had no choice but to set it down on the Bakersfield Freeway directly below him. Luckily, traffic was light. He came in over one car, and as he landed uphill, he nearly caught up with another.

Gene cut off the fuel mixture, slowing the plane. Coming up to the Lancaster Freeway turn-off, he made the turn. This was wise as there is yet little traffic on the Lancaster Freeway and he was less likely to have a wreck and hurt someone.

Braking the plane near an "emergency parking only" sign, Gene hitched a ride to a telephone to call me. The photographer stayed with the plane, taking pictures.

I said, "Gene, you must return to the plane right away and move it off the freeway."

"Oh, it's okay, Claire, it's on an off-ramp and it's out of everyone's way."

"You are going to have to get it clear off the highway because the police are going to be there any minute and they'll get all upset. Put it in a field or something. I'll send a crew out and we'll see what's wrong."

I have my mechanic take our covered truck, with its tools, parts, compressor and generator, and I send another employee with him to help repair the airplane.

Gene hurried back to the plane and found it surrounded by several police units. They had called a tow truck, otherwise my friend would have rolled the plane down into the field by hand. He holds an A&P (Aircraft and Power) rating and he makes sure the plane is not damaged while being towed.

The tow truck pulled the plane down the Lancaster Freeway to an off-ramp, brought it down and rolled it into a field. This area has little traffic; it is a dead-end area. No housing tracts yet. It is safe for now.

The CHP (California State Highway Patrolmen) are upset about the landing, but not the deputy sheriffs and the local police; they want to block the traffic and let my friend take the plane off from the Lancaster Freeway. The CHP says no.

My crew arrived and Gene opened the cowling to look at the engine. As he did so, a bolt dropped out. This bolt is part of a throttle linkage, and when it comes loose, the engine can go either into full throttle or into idle. This time it went into idle. Gene picked up the bolt and put the linkage together. The crew checked it over and ran up the engine.

The CHP refused to allow the plane to take off from the freeway without a permit from the Department of Public Highways. Without it the wings are to be taken off and the plane trucked home.

Gene called me again. I tell him not to worry, that I am coming right away with my son Mike and the crew could go home then. I'd figure out something.

Mike and I load the car with food, water, pillows and blankets. I also take my .38 revolver and wad-cutter ammunition.

We arrive to find the crew has checked the freeway for any obstacles that might endanger the plane in taking off. They tell me there will be no problem in taking off the next morning. I can block off traffic and take off towards the southwest, against traffic. They think this because of the wind direction this evening. I do not say much for I know there is no wind at dawn. I can take off with the traffic, except there should be no traffic for those few minutes. I have this plan....

My crew declines my dinner offer and I send everyone home after listening to the story again. Gene suffers from some dehydration and sunburn but he is a good sport and considers the incident another adventure in his life. The photographer has had a good time shooting everything. We agree to shoot his aerials tomorrow.

After dinner at a cafe, Mike and I drive back for our bird watching. As we make up our beds in the plane and the car, the police come by and ask if the plane can fly at night. "Yes."

"You do fly at night, don't you?" I say yes.

"There is no traffic at night, you know."

"I know." But I'm not about to tell them my plan. I want no witnesses who may file reports.

Mike and I spend a restless night. I moved the divided rear seat of the plane to a reclining position and it is fairly comfortable. Mike slept there. I curled up in the back of my car. It is anything but comfortable. My king-size bed at home is just right for me; the car is a sad substitute. I put the gun and the ammunition near me. I do not load the gun; I can load and fire six rounds in about ten seconds. If it were loaded I might accidentally fire it and hurt someone.

Now to the problem. Getting a permit will take three or four days, with increasing possibility of theft or vandalism to the plane. And the result? A dawn flight with police blocking traffic. Dawn, because in these canyons any other time of day can mean unexpected and dangerous gusts of winds. And at dawn there is little traffic.

Occasionally I hear a truck or something go by and I check the airplane and the area carefully from inside my car. No one bothers us. We are hard to see from the highway unless someone is looking for us.

In predawn light, a car coming off the Lancaster Freeway 50 yards away wakes me. I wake Mike. We drive the car up the on-ramp, then down the freeway eastward to check for wires, signs, bridges, any other obstructions. We find that the freeway has a slight rise in it and after that it goes around a hill.

Beyond the hill is a slight downhill area, then a straight stretch about two miles long of completed freeway before it ends and becomes the regular road to Lancaster. There is enough room with no obstructions for a safe take off.

We push the four-place, high-wing, tricycle-gear Cessna from the field onto the paved side road. I check over the plane thoroughly, going all around it, draining the sump of water collected overnight, and checking the oil level. I climb in, buckle the seat belt, set the brakes and turn the key to start. The propeller winds over, the engine coughs and steadies to a roar. Mike drives the car to the edge of the freeway and gets out.

As the engine warms I check instruments, magnetos, and free movement of the controls. I release the brakes, push the throttle forward and taxi the plane up to the freeway on-ramp sign. I check the side road and the freeway section I can see for the telltale lights of police cars. There are no cars in sight.

Mike stands beside the car. He can see nearly a mile in the direction of on-coming traffic. If there is no traffic, he will wave. He waves.

I give it more throttle. The roar increases and I move the plane to the top of the on-ramp, clearing the yellow "merging traffic" sign. I glance back at the open stretch of freeway behind me. Nothing is moving. Mike waves again.

The diffused light of dawn flows pink and golden down the sides of the rolling hills. Cloudless and calm, it is a beautiful day to fly.

The moment is now if I am to take off before a patrol car appears. I push the throttle forward and move the plane up to the rise in the highway. It gathers speed and moves swiftly around the hill. I watch closely, staying just the right distance from the embankment on my left and the hill on my right.

As we come around the hill on the downslope, the plane reaches flying speed and I pull back gently on the wheel. The plane lifts off as the sun tops the hills. I gain altitude, make a one hundred and eighty-degree turn and fly back up the highway. Mike is standing beside the car a short distance away. He had followed me around the hill.

I waggle the wings in a "goodbye, see you later," and turn towards home, Santa Monica Airport.

That afternoon I get a phone call from a deputy sheriff. He had been flying a helicopter the previous day on a fire call in the Newhall area, hadn't had a chance to come over to see the airplane and wants to know if my friend got it back into flying condition.

I say, yes, everything is fine. He says he hopes to get out to see the plane. "Well, it's back here at Santa Monica."

"Oh. How did you get it back there?"

"Well, I took a crew out there in the middle of the night, took the wings off, and just as we were getting everything into our big truck about dawn, an old rancher who lived down the highway a few miles, stopped to see what we were doing. He was the sweetest old thing. I don't remember his name but he was a regular doll. He said he had a hanger and a landing strip on his ranch. We were welcome to go there and put the plane back together instead of risking damage to it by hauling it clear back to Los Angeles. So we took him up on it. We went over there, put the plane back together and I flew it off his private strip."

"Oh, that was nice." Then the deputy says, "There's no traffic on the freeway at that time of morning, why didn't you just take off on it?"

But, it is illegal to take off on a freeway. I wasn't about to admit to authorities that I had done so without a permit.

A commercial pilot studying for his instructor's certificate flew the photographer back to the area. We had no further trouble with that plane. The incident gives us a fun story.

My only fear in taking off on the freeway was that the CHP would see it, arrest and fine me for doing so without a permit and official supervision. They are the ones who care.

Safety is their first consideration, as it is mine. I gambled that at dawn the CHP officers would be in a cafe having coffee. Even if they did see me, a fine was much cheaper than taking the wings off, and with much less delay than in trying to get a permit. This was a work plane; I needed it back in service and I needed to get back to my office.

Back in the office, I record the flight data in my current logbook. It is nearly filled.

"Two SM Pilots Fly Mexico Mercy Hop" read a Santa Monica newspaper headline. The story gave only a few facts. It did not tell the story of terror, hostages, intrigue, corruption, and the payoffs it took to rescue Harriet Hagedon.

Early in 1962, Hank Gammell, chief pilot for Motorola at Santa Monica, phoned me to say that the ex-wife of a friend was in deep trouble in Acapulco, Mexico.

Two friends, Joe and Pat Colvin of Colvin Nursery in Malibu, had taken an airliner to Acapulco to visit Harriet, planning to drive back to California with her in her car.

They found Harriet in a hospital with hepatitis and under armed guard. Apparently, Harriet, an "Aunt Mame" type, had failed to renew her visa. This was discovered by the hospital and she was being held hostage, to be ransomed by relatives.

Joe and Pat Colvin had spent all their cash trying to get her out of trouble, but they may have started too low on the totem pole in making payoffs. They called home for help. They were able to get her out of the hospital and back into her apartment.

Hank Gammell asked me to take his twin-engine Piper Apache and bring her home. Going with me was my then partner, Betty Faux, and Harriet's sister, Ethel Beck. We took off for Mexico, with our first stop at Mexicali to get a tourist visa for Ethel.

Ethel had been apprehensive about flying but soon became amused and relaxed when she saw how nonchalant Betty Faux and I were, drinking coffee, eating fruit, crackers, cheese. She became our stewardess and also took pictures of the sea coast, a small mountain village, and a smoking volcano.

We landed at Hermosillo for fuel, and finally, at Mazatlan to spend the night.

I phoned Acapulco. Joe said, "Come on down, but it doesn't look like any of us are going to get out of Mexico. Things aren't going well."

Chilling words. Among Hank Gammell's last instructions to me were, "If anything goes wrong, burn the plane, Claire." How could I do that from jail? I was carrying three bundles of pay-off money. With sinking hearts, we went to dinner.

With our early-morning takeoff, we arrived at Acapulco at noon. Ethel phoned the apartment while I refueled the plane. Joe said that no progress had been made and that they were all under armed guard. "Come on in."

We took a cab to the apartment. Inside were several uniformed men—local police, Immigration, Customs, and other authorities, all involved in the negotiations, all with their hands out. They made many phone calls to Mexico City with always some special charge we had to pay. Joe kept stuffing money in all their pockets.

By late afternoon it appeared that we were nearing the end. Betty Faux and I went to the airport to pay off El Commandant. It was agreed that we could leave Mexico as long as we could get out without spending the night anywhere because of Harriet's hepatitis.

El Commandant cleared us through to San Diego, for another charge. I wanted to take off right then, but there was no night refueling anywhere on our route. Then I suggested we take off at four a.m. so that we could make Puerto Vallarta at seven a.m., in daylight. That was the answer and everyone began to relax.

Then the commander said, "Our tower is not open at that hour."

I told him I didn't need the tower; there was no other traffic at that hour. He insisted. It cost another $20 for the tower.

He said that the lights would not be on. I told him I didn't need the lights. He said I did. Another twenty to end that discussion.

Finally, it was done. The flight plan was filed, and messages sent up the line to our intended stops for fuel so as not to delay us. There was more paper work to do and the tension and worry continued until seven p.m. when we were able to return to town. All this time we were trailed by two uniformed men.

We all went to the airport with Ethel and Harriet at three a.m., pre-flighted the plane, loaded the two passengers into the back seat where they could sleep most of the way.

Betty Faux climbed into the left seat, and I, into the right. The runway and taxiway lights came on, the Tower cleared us to taxi out to the take off line; we did the engine run-up, then the Tower cleared us to take off.

I had Betty fly straight to the coastline. I taught her to fly and we flew together in a Powder Puff Derby, but I could never quite stop instructing her even though she was a great pilot.

When I spotted the white tops of the waves, I had her turn right and stay off the shoreline enough for me to keep them in sight. That kept us clear of the Sierra Madre mountains. Mexico has limited navigation aids and no lights at all over much of it. Betty flew on instruments for over an hour as that land at night is black.

It began to get light and Puerto Vallarta was a welcome sight at seven a.m. Refueling took an hour. Gas men poured the fuel from large drums into a bucket, then that fuel into another bucket covered with a chamois for a filter. This was duplicated at Mazatlan and at Hermosillo.

We cleared customs at San Diego and arrived at Santa Monica at nearly seven that evening. Harriet was rushed to a hospital.

Pat and Joe Colvin were stopped by the police while driving Harriet's car out of Mexico. They accused Pat of being Harriet. Finally, they were convinced otherwise and let the pair go.

Harriet's escape cost $22,000, most of which we had carried into Mexico. We had flown a 3500-mile round trip in three days. We would never forget that adventure.

Chapter 13

Soon after I opened Claire Walters Flight Academy, I became acquainted with Peter Ahrens. He was employed by Avis of Australia, and eventually became president of the rent-a-plane division. His job was to buy airplanes in the United States and get them to Australia.

Ahrens disassembled and crated the single-engine planes for shipment to Australia; the twin-engines he flew there. I met Peter during one of the times he was preparing twin-engine planes at Santa Monica. We didn't become close friends until he brought Rosalind Merrifield from Australia to ferry a plane. They made my offices their headquarters and we were able to spend time together.

Roz was born in Sydney; Peter immigrated from Sweden. She had a pleasant Australian accent; Peter, especially when he was agitated, yelled in half-Swedish, half-French, half heavens knows what else, and was hard to understand.

Roz was an accomplished ferry pilot and navigator, flying from various countries to Australia many times. She had also flown with a geophysical survey group on Guadalcanal.

Avis wanted Roz to ferry a plane from the United States to Australia for its publicity value—she would be the first Australian woman pilot to do it, and this would be entered into the record books.

In the previous year, 1965, Betty Miller had flown it, becoming the first American woman pilot to do it. In 1964, Joan Merriam Smith, flying solo, completed the first around-the-world flight by an American woman, in her ancient Piper Apache.

I was able to make a coup for our Long Beach, California, Chapter of The Ninety-Nines, by having Roz Merrifield and Joan Smith join it. Miller was a member, as was Betty Faux. We then had all the transpacific-flying woman pilots in our Chapter.

Roz worked on her American flight ratings while she was at Santa Monica waiting for the planes to be ready. Betty Faux gave her an instrument rating and later, when I became a designated FAA examiner, I gave her the commercial license.

In Australia she had all of these, plus an instructor's rating. She'd logged 1000 hours instructing.

Back in Santa Monica to prepare for yet another flight to Australia, Roz and Peter asked Betty Faux if she would like to fly a plane to Australia. They didn't ask me as they thought I couldn't leave my business or my children, Michael and Susan. I never mentioned to Pete or Roz that I dreamed of such a flight.

While on that flight, Betty told them that I would like to go. On their return to Santa Monica, they asked me.

Peter busied himself buying aircraft, preparing them for flight or for shipment, checking on those in overhaul, ordering special fuel tanks for the three we were to fly.

I started to learn celestial navigation and bought approach charts for Honolulu, Canton Island, Fiji, so forth, in case we had to make instrument approaches. Fortunately, all our flights into these places were VFR, and we flew only during daylight hours. Oh, well, no learning is ever really wasted.

Roz and Pete worked with me on flight charts, plotting our trip. Pete actually worked out the headings we used, but I worked on them myself, made notes, and discussed them with him.

I told everyone I was going and many friends came to see if they could help. Ralph Thomas of the FAA told me that the thing about an overwater flight was not to think of it as being over water, but to think of it as being over Kansas snowfields. One can think of it this way because most of the ocean is covered with clouds most of the time.

Today, in the 1990s, a wonderful navigation instrument is available. The GPS (Global Positioning System), small enough to be held in one hand, tells the pilot where the plane is within 50 feet. Satellites over our planet enables this tracking system.

The GPS enabled balloon racing from North America over the Atlantic Ocean to Europe, a balloon flight from Japan to eastern Canada, many small planes in formal races around the world, and much safer plane flights over comparatively unmapped places like most of Africa and the vast Pacific Ocean. This makes navigation easy — as long as that particular instrument works.

The rest of the requirements for flying safely over water remains the same: A competent pilot with common-sense gained through real-time flight-time, pilot's ability to take off with a ten percent to 26 percent overload and far out of the center of gravity, fuel tanks not leaking, pilot's know-how in switching fuel lines to balance the plane, keeping the plane on step and at a normal 2400 rpm, weather knowledge, a working radar, working radios with competent knowledge of them, and engine(s) operating properly.

In the 1960s, navigation over water by airplanes was pretty much the same, except for radio aids, as it was for centuries in sailing craft.

Celestial navigation during the night and sun shots during the day. Knowledge of wind directions and velocity aided the aviator, just as ocean currents and their velocities and wind drift aided the sailor in his navigation.

Navigation to tiny places like Canton Island, three-and-a-half square miles and nine feet high, has to be precise. Peter could do it for all of us with sunline checks. I had a sextant and practiced with it, but I didn't trust myself and left it behind.

Peter and Roz were to fly Piper Aztec Bs, while I would fly an Aztec A. I'd actually had only about 30 minutes in one as a pilot. Some years ago I had instructed a man in his ship, and I gave others instruction in that type, including Betty Faux. We rented one and Betty had about two hours in it before leaving for her transpacific flight.

When Peter bought the planes for us to deliver to Australia, he gave me a multi-engine check ride, which was a good idea since I hadn't had one in 17 years. I had several hundred hours of multi-engine time, but the one I had taken my check ride in was a UC-78, the "bamboo bomber", in the 1940s.

I took off in the Aztec with Peter in the right-hand seat. He said that in my first single-engine procedure I would have killed us, but he was exaggerating considerably. After that, everything went well. The Piper Aztec is an easy airplane to fly, much easier than most twins. It is a fine single-engine airplane should one quit in flight, and a fine one for women as it is so easy to control.

The three planes were completely overhauled and every piece in the paper trail signed off properly by the person certifying it. Each piece of overhauled equipment was tagged with the serial number on it, and all other information required, following the safety system required by Australia.

Peter Ahrens, when getting an export license in the United States, "pulled down" the aircraft to be flown across and while they were down for overhaul and inspection, had all modifications made that were necessary to meet the Australian regulations. He had all the air navigation orders that were applicable in Australia outlined for each different kind of aircraft, and all of these were completed and properly signed off before each aircraft was reassembled. This saved a tremendous amount of time and money later in Australia besides insuring a safer transpacific flight.

New aircraft direct from the factory, with brand-new Certificates of Airworthiness, were automatically exportable. Distributors in Australia have their own shops and are set up to make the few modifications necessary to the new aircraft.

Pete bought used aircraft and each presented its own set of problems. His twin-engine aircraft were nearly ready to be placed into general service on arrival. The extra fuel tanks had to be removed and the seats reinstalled.

Pete and Roz were careful to fly these planes properly during their transpacific flights; the planes would be flown by them and by other Avis employees in Australia. Some of the other ferry pilots they knew were flying them well under the proper rpm in order to save money on fuel, and many of these planes had engine failures within 20 hours after arrival in Australia and had to be overhauled. We have no figures on how many failed during the flights, for these disappeared forever.

Peter had extra fuel tanks built in Hillsboro, Oregon, to his specifications. These were metal tanks, not the rubber tanks put inside of crates as was the practice of some other ferry pilots.

Three tanks were put in the fuselage itself and carried 400 gallons total in addition to the 144 gallons in the wings. The three tanks were called rear tank, front tank, and top tank. The top tank carried less fuel than the two bottom tanks.

Each tank was rigged so that it would feed directly to the engines. The fuel tanks had lines with a series of valves to be turned on and off when changing tanks. This extra fuel in the fuselage placed the plane far out of the center of gravity.

In the past, the FAA had allowed airplanes to be only ten percent overweight. These Aztecs were 26 percent overweight and eight inches out of the center of gravity. This meant the weight was towards the rear. Planes will fly under these conditions...I will say they will fly...the only thing is, if the tail-heavy airplane stalls there is no recovery.

Because of this extra weight, these tanks had to be put as far forward as possible. That meant that the pilot's seat (the left-hand seat) had to be as far forward as possible, leaving just enough room for the pilot's knees. The mechanics took the back of the left seat off, removed the entire right-hand seat and the right-hand set of flight controls.

The gas tanks were placed as far forward as possible, right behind and against the pilot's seat. Each of us had to sit leaning back against the tank itself.

Now the planes were ready for the final equipment loading and fueling. Roz called me from Hillsboro and said to come on up.

I kissed my children goodbye and took a United Airlines flight from Los Angeles International Airport to Portland, Oregon. It was a clear, bright day and when the Boeing 727 was at its assigned altitude of 35,900 feet, I could see the magnificent Sierra Nevada,

Lake Tahoe, and later, Mount Lassen, Mount Shasta, and just as we started down, I saw Mount Rainier far north of us where it rises so high near Seattle, Washington.

We landed at Portland, Oregon, and an hour later Roz found me. Peter was waiting in an Aerocommander.

At Hillsboro, Pete and Roz showed me what had been done to ready the aircraft for our transpacific flight. I peppered Roz with questions about every aspect of the flight, things I hadn't had time to think about until then.

It is Sunday, June 5, 1966, ten a.m. We three drive out to the field and the three planes. Peter gives us our survival gear: each plane is to carry flares, flashlight, one Mae West (inflatable life jacket), two microphones and two headsets, an empty jug apiece for Roz and me (Pete has his own system), water bottles, a Regency Standby radio, survival book, oxygen bottle, first-aid kit, hunting knife, handaxe, and a two-man life raft.

Peter and Roz each have a .38 pistol (for sharks). I don't know anything about guns of any kind so I won't take one. I figured with my luck, instead of hitting a shark, I'd just shoot a hole in the raft.

After I returned from Australia I took a course in pistol shooting at the Culver City Police Department. They held a women's class each month and I eventually became an instructor. If there is another transpacific ferry flight for me I will take a pistol.

After breakfast, we are photographed and interviewed by the local news people. Pete holds a survival class for my benefit, and the news folks sit in on it. This is okay with Pete — he loves an audience, and goes into greater detail than usual.

Besides survival, he discusses the take off with overload, the fuel system, the oxygen, and related subjects.

The survival course is quite thorough. Pete tells me how to land in the ocean, how to get out of the plane, what equipment to take, and what to do if I am lucky enough to float to an island. I listen; I am a poor swimmer afraid of the ocean.

Pete says there is only one way into an island. There is always a hole to the lagoon through the coral reef and it seems that one will float around an island following a current and must plan to get through that hole while at the same time staying away from the sharp and dangerous coral itself. It can rip the raft, and if my shoes aren't sturdy enough, the coral will cut my feet, which will result in a dangerous infection.

Many South Seas islands are small sand and coral islands just barely above sea level. If there are coconut trees, one can survive a long time. Coconuts are one of the finest of basic foods, providing both food and liquid. Natives carried them on long sea voyages. Some of the fish can be eaten, but many are poisonous. I'd have to read the manual and hope it is right.

But before I can climb into a life raft, I have to land the aircraft properly in the water, Pete says. It is necessary to land cross-wind in between the swells. The plane may skip across the water and then settle, if I'm lucky, into the ocean and sit there on top, but perhaps well into the water itself, depending on how empty the tanks are. Possibly the tail is torn off in hitting the water.

Then, if I'm still lucky, I can open the door and climb out on the wing, reach back in for the life raft, drag it out and inflate it after attaching it by rope to the door of the plane.

My life jacket would be on me from the start of the flight, and, again, I'd have to be careful not to inflate it until I was in the life raft.

This will take a lot of thinking and planning of what not to do as well as what and when to do each thing as fast as I can. I have never had any practice in ocean survival technique and there is no time now.

After the life raft is inflated, I must get my shoes off, put them in the raft, unload the survival case containing food, flares, a first-aid kit, and some tools, putting them into the raft. The box itself is too heavy when loaded for me.

I am to throw the half-empty plastic water bottles (we had partially emptied them when we bought them) into the water to float until I can pick them up later. The emergency radio set is next to last to go into the raft.

The last is me. I am to climb into the raft in my bare feet so the bottom won't get ripped; I must untie the line quickly as the plane is sinking by this time.

If the door cannot be opened, Peter says during his lecture, then I am to take the handaxe and knock the windshield out. Now, the windshield in an Aztec is small and it will be quite a project to get myself and all that equipment out, but I suppose fear and the rush of adrenaline might help me.

The emergency radio set works by pulling out the antenna. The set sends out a signal which, theoretically, is picked up by the rescue plane. The flares are a help if the rescue plane arrives at night. We also have a shark repellent.

Two or more planes flying as a group gives the downed pilot a greater chance of survival. The downed plane's position would be sent and a rescue possibly made. It's a big ocean; finding someone is difficult unless the raft is kept pinpointed.

We arrange that the two planes still in the air will circle the fallen aircraft to see if the pilot gets out. If the pilot doesn't get out and it sinks there won't be any need to send for a rescue plane or ship, risking others just to bring back a body. If the pilot survives, the position is noted and a Coast Guard rescue may be possible.

Class over, Peter flies his plane to make a final check on his radio sets. After he is through, all three of us take off and fly in formation to the sea coast where photographers in an Aerocommander photograph us in formation. An hour and 12 minutes later we taxi back in, only to find that one of Roz's cabin tank fuel lines is leaking badly underneath the fuselage. Pete said it had to be fixed that night.

After dinner, Pete and I drop Roz at the motel and we return to the airport. I work on my charts until he calls me to help. I climb into the cockpit and flip on the master switch, then the booster pumps, each one at his specific order, then the main tanks (inboard) ON then OFF, then the front cabin tank ON. Pete, with relief, finds the leak is easily fixed.

I pack all the rest of my survival gear in the plane, check that the two-man life raft is secure on top of the top tank, grab my charts and we drive back to the motel.

Most of Monday, June 6, is spent in indecision: to fly or not to fly. First the weather, then the radios. Finally, we leave Hillsboro at four-thirty in the afternoon, winding our way around the storm by flying over the sea near the coast. Four hours later we land in the rain at Oakland. We carry fuel in the wing tanks only; we will fill all the tanks in Oakland.

Tuesday, June 7. We partially fuel the planes, filling rear tanks and top tanks. Oil is brought up to the proper level. My HF set is repaired. Parts for Pete's all-weather radar have not arrived; he will have to fix it in Honolulu. He fixes a small oil leak on my right engine.

At a supermarket we buy food for our baskets. Nearby is a store where we each buy foam rubber to put between our backs and the gas tank. A 14- to 16-hour flight lay ahead; an aching back will not help.

Tonight we pack our styrofoam baskets. We make our sandwiches, ham and cheese, cornbeef, so forth, just like we like them. Cans of baby food on the bottom, then sandwiches, crackers, potato chips, candy, oranges, apples, dried fruit and canned fruit, fresh strawberries and cherry tomatoes. This is always the best way.

The prepared lunch boxes we bought later were pretty sad; here we fill the boxes with food we like.

After I pack the food, I pack my clothes into two cases: items I will carry with me are in one small bag, while the other is to be shipped on Quantas Airlines with Peter's and Roz's bags.

Pete is touchy about weight and insists on checking my suitcase for unneeded items.

Wednesday, June 8, three-fifteen a.m. Up to get ready. By four we are at the field readying the planes for the flight. I am to oversee the refueling while Pete goes to the weather bureau. He returns just as the fuel truck arrives.

I put my suitcase, the camera case with both a 16mm movie camera and a 35mm still camera, and the food basket on top of the emergency equipment on the front right-hand side where the seat has been removed. The seats and other removed equipment have been shipped to Australia.

The cabin door is on the right-hand side, which means I must climb over all this stuff to reach my seat. In an emergency I'm not sure I can get out.

Some long-distance flyers remove the left seat and put all the equipment on that side. The major drawback to this is that in instrument weather, the pilot is at a serious disadvantage trying to read the one set of instruments in place on the left-hand side.

Each plane has a complete unit for navigation, provided everything works. Our radio equipment consisted of an ADF, dual VHF navigation-communication sets, and an HF set. On the flight, sadly, only one of our three HF sets worked, even though the others were worked on in Oakland, Honolulu, and Canton Island.

I drain the sumps and check the oil. My front tank and wing tanks are filled by the fuel attendant. Roz's plane is next to be fueled and then Pete's.

The seam of Pete's front tank cracks and gas streams to the ground. The day is spent pumping out the gas, taking the tank out, fixing the tank, putting the tank back in and refilling it.

The fuel leaks and cracked tank occurs because of the high pressure hose used by the gas men. Normally, Pete would fill the tanks or have us do it. But some places won't let you do it. There's no problem when light weight hoses are used. One can build tanks only so strong before their weight becomes too much.

At Oakland, the man filling the tank uses a high-speed pump as he knows we'd be using about 600 gallons per aircraft, and he gets tired and cold standing out there at four a.m.

The fuel is pumped in at a tremendous velocity and it breaks the tank. The fuel men have been told this so they know better, but these gas jockeys, doing a midnight to dawn shift, just don't care; there are so many planes coming into Oakland.

After dinner that night, Pete and I work on a new flight plan. This morning, a headwind called for a 16-hour flight. We hope for better winds tomorrow and that the stratus cloud overcast will clear earlier than it had today.

Chapter 14

Thursday, June 9, 1966, four a.m. Oakland, California, to Honolulu, Hawaii. Today is the day. I am up, dressed, packed, and otherwise ready when Peter knocks at the motel door at five. The weather office has told him there is a large hole in the overcast at Alameda. To fly VFR, we will have to stay below 1000 feet until we reach there. Visibility is good.

We drive to the airport, file our flight plan and walk to our aircraft. After we have the tanks topped, Pete discovers a small leak under Roz's plane. He fixes it, then climbs into his plane and fires up the engines. Roz starts hers, and I try to start mine. I keep trying but with no success.

Pete shuts down his engines and comes to help. We get mine started, only to see a mad stream of fuel pouring from his plane. He fixes it, somehow, in seconds, and we taxi out to the runway.

Seven-twenty a.m. I turn on my radio and tune in OAK ground control. I receive nothing and discover that my VHF transmitter switch is fouled up. I must switch to one position to transmit and to another position to receive. I decide to put up with this inconvenience rather than cause another delay.

Ground control has us taxi to Oakland's shorter runway. Oh, no! Peter had assured me we'd use the long one. Peter!?!

Pete moves down the runway, lifts off near the end, then Roz moves down it. I wait until she is off before I gently move the throttles forward. As the ungainly beast gathers speed it slithers, bucks, and wallows back and forth.

Pete had told me to reach 128 miles per hour before lifting off, but as I hit 122 mph, I ease back on the controls — the end of the runway is only yards away. It lifts off and a few moments later I hit the button for gear UP. I smell burning rubber as the wheels slide into their slots. What a burden it was on those poor tires to carry all that weight! We have 2090 miles to fly and 15 hours before they must roll again; the plane will be a lot lighter.

After spotting Pete and Roz, I look at San Francisco, brilliant in the sunshine striking it between the clouds. A few minutes later I take a couple of pictures despite the plane's wallowing.

Soon after we left Oakland, Pete took an ADF (automatic direction finder) bearing off of the Farallon Islands transmitter.

Before take off he had figured our heading correction for the winds given him by the weather bureau. After getting the ADF bearing, Pete shot the sun with his sextant to get a bearing on our distance from San Francisco to the Coast Guard ship November stationed halfway between Oakland and Honolulu.

November maintains communication and navigation facilities as well as performing rescues. I wish they had a dozen like her anchored along our path.

The Farallon Islands form a national wildlife refuge with a radio transmitter station on its largest island. It is directly on our course, 30 miles out from the Golden Gate Bridge.

I know that as the sun gets nearer the horizon, it becomes harder to determine our position. If the radio equipment goes out, it might become difficult to make a landfall. It is important to make frequent navigation checks. From this we can also tell what our wind drift is and we can make slight compass adjustments. Our planned flight of 15 hours will bring us into Honolulu at sunset. We are racing the sun.

Nine a.m. We are an hour past the Farallon Islands. Pete and Roz are having difficulty transmitting our position on their HFs to Oakland Control. Pete asks me to try.

We must make frequent reports because authorities want to know where we are. If they don't hear from us they have to think about making a search. And if we just haven't been calling, they get upset.

I switch the transmitter knob to the position needed to transmit. I answer Pete, then switch the knob to the HF position. The knob comes loose in my hand. The VHF set is stuck on the receiving position. I can hear Roz and Pete but I can not talk to them.

The two discuss this at length while all I can do is listen. Pete: "But I gave Claire the frequency we are using!"

For the rest of the flight I am unable to speak by radio. I can receive Pete and Roz clearly but I can not respond. Unfortunately, we have not set up alternative signals.

We fly in a diagonal line so Pete, in the lead plane, can look out of his left-hand window and see us. Roz is on his left and I am on her left. It is hard to keep leaning forward to keep them in sight through the right-hand side of the windshield.

We can't fly really close together because we each have things to do. Peter has to keep shooting the sun, we have to work the radios, later on we'd eat, and I frequently make notes and take pictures. Flying close together would be dangerous.

Those who have driven in car caravans know the problem. In getting through traffic, through traffic lights, so forth, it is difficult to stay together. The first car can maintain, say, an even 30 miles an

This Flying Life

hour, the second car will have to do 40, and the third car, 50 at times, to keep up.

In aircraft you get these slowups, you get out of position, you get off the step, you feel that you want to adjust the mixture controls because the barometric pressure is changing and you don't want to run it too lean or too rich — you'll get overheating on either, you know — so you check your engines every now and then.

Roz told me some months ago as we were planning my flight with them, that in the previous year, alone, 19 aircraft went down between San Francisco and Honolulu, most of them flown by professional pilots ferrying planes. One was not; he was a medical doctor in his Piper Aztec twin-engine, or maybe it was a Cessna 310, who wanted to fly to Honolulu but wouldn't take the time to prepare properly.

He knew he needed extra fuel so he added some tanks. He expected to have Omni all the way, or at least, ADFs, or something, but he didn't bother finding out what. When he realized that he had two hours of reserve fuel and he wasn't getting any Hawaiian ADVs, he called Honolulu and said, "I've got two hours of reserve fuel, can you help me?"

Honolulu said, "Start transmitting and we'll have our aircraft and trackers bring you in."

An aircraft carrier was moving along 600 miles north of Honolulu. Its radio operator listened to this exchange, called Honolulu and said, "Look, I know this is pretty ridiculous, but I think I'm getting this guy."

The aircraft carrier command had the pilot do a few turns and picked him up on radar. He couldn't have picked a better place. Of course, it was the only place in 600 miles.

The carrier took control, got on the VHF radio with him, turned the carrier, lined it up to cut the swells, told him where to land, how to land with the gear up, got their helicopters off and waited. He landed, ditched his plane perfectly, and jumped up on the roof where a helicopter picked him up. He hardly got his shoes wet!

Roz told me about a fellow countryman who had prepared his aircraft but not himself. He took off from San Francisco twice and turned back twice, panicking. No weather or engine problems, just panic.

On the third try he continued his flight to his point of no return, decided that he couldn't go back, that it wasn't safe enough. He found that his long-range radio wasn't working and asked a Northwest Airlines pilot to transmit for him. He decided to detour to an alternate airport. But unless there is an aircraft carrier out there just waiting, there are no alternates.

Honolulu radio said, "Where do you want to go?"

The pilot said, "To the Aleutians!" The Aleutians are 2000 miles north of where he was; it was 1000 miles to Hawaii.

Roz said that a Continental Airlines plane came along, picked up his signals on radar, talked to him on the radio and persuaded him to continue on to Honolulu. He was given all his compass headings to the Islands and his ETAs (estimated time of arrivals). They called Hilo and had the Coast Guard come out to meet the plane and guide it in.

This was a supposedly professional pilot. While they might do well in their own territory, with everyone taking care of them, they don't have a clue as to what is going on when they are on their own. They assume — without actually finding out — that they'll have radio aids to follow all the way, that there will be all sorts of traffic going all the time. He survived that time, but too many others ferrying aircraft across the Pacific Ocean that year and the years before that and since then, have not.

As Roz pointed out, the trouble begins well before the start of a long overwater flight. The ability to handle a 26 percent overweight craft that is eight degrees out of center of gravity has to be acquired over many hours of real-time flight. A good pilot learns the feel of a plane, when it's ready to fly, how to handle it on turns and attitudes, takeoffs and landings.

Eventually the pilot acquires the ability to "fly by the seat of the pants"; that is, he knows his aircraft thoroughly, its capabilities and its limitations.

Roz told me of another young fellow that a veteran ferry pilot was helping to get ready to fly a Comanche to Australia. Pete bought the wreck when the fellow crashed it trying to take off from San Francisco International Airport.

The young man decided to take his girl friend along as his "navigator". The professional pilot equipping the plane had given the man a contour seat to sit on but did not know about the woman.

This contour seat was a special gas tank seat which fits around the tank, allowing a bigger fuel tank. The old pilot did not remind the young man how the stalling speed increases with increased weight, assuming that he knew.

The veteran pilot probably had not thought about the need for increased speed for many years; he was so used to adjusting for it that it was automatic for him to do so.

In different types of planes, the center of gravity moves either forward or back with an increase of weight well over the normal maximum. If the c.g. moves way back, the elevators and rudder will be less effective. If the c.g. moves forward, it can be too positive and you will not get it out of its downward direction at "normal" speed.

The Comanche, overloaded, had its c.g. move back. The young man had his girlfriend lie across the top of the tank for the take off. We don't know where she was supposed to sit later.

The young man guided the plane down the runway and took it off at its normal speed. It probably even felt like it wanted to fly; ground effect does that. But it stalled violently, dropped a wing and dived right on in, cartwheeling. The passenger was killed and the pilot was in the hospital for quite a long time.

It's now 2:24 p.m. We are nearing the stationary ship November. We had picked it up with ADF needle at 12:45 p.m. At 1:45 p.m, I heard the code for the first time. Much of the ocean is covered with a stratus deck and our altitude varies from 6000 to 7000 feet to stay clear and above it, so until now I have seen as much of the ocean as one sees going from Long Beach Airport to Catalina Island.

Now there are large breaks and I am hoping to get a picture of the ship. I'll also try to transmit to the ship as I am now uncertain whether my transmitter switch is stuck on VHF or HF.

I have my Aztec on autopilot so I can write in my journal. Cumbersome rolling and pitching of the plane prevented me from doing it earlier.

Pete asks Roz if she thinks I can listen to 131.0. If only we had set up some signals before we took off!

According to our flight plan, we are to be over November at three p.m. Heavy clouds again cover the ocean; I may not get to see the ship.

My friends' compass headings wander between 210 and 240 degrees so I must keep readjusting my autopilot. Whoops, I just went through Roz's propwash! Must stay clear!

I have climbed a little above my friends and they are hard to see against the clouds; I gradually descend. At 2:50 p.m. I switched from the top tank (two hours ten minutes fuel out of it) to the front tank. My tanks have been working well. I used the rear tank first after taking off on the mains.

My ADF points directly towards November. I hear Pete say that we will be there at 3:15 p.m. The ADF needle reverses at 3:20. There is solid stratus now beneath us. I put my camera away. I try to call the ship; no luck.

I hear Pete tell Roz that he can't understand what's happened to my radio (or rather, me) and that he will have a word with me (said in his very best Swedish Father tone) about not staying where he can see me. If they only knew how I would love to talk to them in my very best American accent! And, I can't stay where he can see me all the time because he wanders all over the sky.

Once again I silently thank Ralph Thomas of the FAA office for telling me to just pretend I am flying over the snow covered fields of Kansas and it's all okay!

I ate breakfast after leaving the Farallons and lunch as I approached November. The corned beef and cheese sandwiches are delicious. A few chips, cookies, and small tomatoes add up to a good lunch. The strawberries are spoiled; I seal them in a small plastic bag and place them in the bottom of the box.

We have flown for eight and one-half hours. I've sipped some distilled water, but dare not drink much; I still have a long time to sit.

I miss not having my usual many cups of coffee — I had none today for the obvious reason. I cannot move, crammed in as I am into the small cabin space left in front of the tank. I hear Pete and Roz discuss my ability to answer a nature call because of my flight suit over slacks. He decides I won't bother. He is so right.

It's now 4:45 p.m. Roz says she thinks she has a fuel leak. Peter answers that we can make Honolulu okay but might have to set down at Hilo instead. My left engine is rough — but it's probably just a fouled plug; this has occurred before. The propeller keeps getting out of sync with the right one.

We are at 6000 feet, our heading is 230 degrees, 22.5 minutes, engines running at 2450 rpm, and lean on the fuel mixture. Everything is all right.

Five-thirty p.m. Roz asks Peter if she should run the tank dry, but he can't understand her. Neither can understand the other. I can understand both and can't help them.

Five-forty p.m. We can see the sea again. We have just left a solid undercast; only scattered stratus now. The sea is beautiful, nearly calm. We have flown for ten hours.

Pete and Roz have discussed many times the reason for my silence, and I wish we had chosen a standard communications frequency such as 122.9. Then I could use the standby set.

Whenever I hear them say they are switching to 121.5 I try to call, but to no avail. The frequency chosen to communicate to one another is 123.45! Now I think the knob on my standby set is bad, or maybe the battery in it is dead, too.

Roz decides to find out if I can hear them before it gets too dark. She asks me to move up close to her and waggle my wings if this is so. I do, happily. "Yes, yes!" calls a delighted Roz.

We start down through beautiful cloud canyons, lit a soft pink at the tops of the clouds by the late afternoon sun. Honolulu approach control is now in contact with Avis Flight One, Two, Three, and we are told to report at the TUNA Intersection, a reporting point on an approach to the Islands.

A little later we are asked to confirm that we are north of our course. A long silence. Then Peter confirms that we are 60 miles north. The sun is too low on the horizon for him to use his sextant for a sun shot, so he establishes our exact position with radio bearings off Oahu.

Nine-thirty p.m. It's still June 9, but it feels like I've been sitting here forever. One hour more to Honolulu. A Navy jet flight appears and flies near us. We must be close to their practice area. Sure enough, when we report again we are told to call when we get on the ground.

We wend our way around tall cumulus clouds, approach Molokai, I see its beacon, and then, suddenly, there are the beautiful lights of Honolulu. The sun has set and the last glow of twilight fades from the sky as we each in turn approach for the landing.

I am still using fuel from my cabin tanks so now I switch to the inboard mains (my wing tanks are all but full), lower my gear and - S-turn to stay behind them. But, they are also S-turning, and I'm getting too close, so I make a 360-degree turn on final. I make the bank too steep and even though the cabin tanks are almost empty, the ailerons feel sloppy and it is extremely difficult to roll out of the turn.

I make a good landing and roll out to the end following Roz. I have trouble getting my feet on the brakes without my right knee hitting the throttles.

We taxi to Joe Jones Pacific Flight Service where we are helped out in style, with leis around our necks and ceremonial kisses, at least for Roz and me.

We unload our gear. The "other" Claire Wolters appears and we meet at long last. He had air shows at Chino, California, and people would tell me about him. Veteran ferry pilot Max Conrad, just arriving, also, comes over and we shake hands. Someone tells us that the Coast Guard has gone out to help Sheila Scott, flying in her Piper Comanche 260 from Australia. She has radio trouble and needs assistance.

Sheila is on her around-the-world flight — the first British woman to do it. I want to stay to meet her, but a 15-hour flight without moving one's legs is slightly paralyzing and tiring to say the least. Maybe we can see her tomorrow. But, as it turned out, I am not to meet her for many years. Then she was a guest in my home for a month in the 1980s.

June 10, five a.m. We are staying at a nice hotel on the beach. My head aches from the aircraft radio noise from yesterday; I had worn headphones for the entire flight.

All the noises of a busy city now crash in on me — loud radios, cars racing in the streets, boys shouting, Navy aircraft, and even what should be beautiful sounds, wind and surf, only adds to the din. Obviously, no one goes to bed in Honolulu or stops partying. My own voice sounds funny to me and I do not feel well.

I struggle out to the balcony to watch the sunrise and the surfers at Waikiki. I take several pictures with my still camera before I flop back in bed for a couple of hours.

After lunch, we three went to the airport to see about our planes. Roz and I wrote messages on Sheila's plane (it's covered with autographs like Joan Merriam Smith's was) and took some pictures. Later, after phoning home to my children, I wandered through the nearby market shopping for souvenirs to mail home.

Saturday, June 11. Peter calls to say he and Roz are going to the airport, that my generator needs work — brushes are worn on one of them, but other than that my plane is ready to go. All the knobs were fixed. We'll be leaving for Canton Island tomorrow. All I've seen of Oahu so far was the airport, Waikiki Beach (from my balcony) and the market. Oh, well, perhaps on the airline flight home I'll stop for a few days.

I cram all the gifts into a small box and take it to the post office. I send a telegram to the Newporter Inn at Newport, California. The 25th anniversary of the Santa Ana High School graduates of 1941 was being held that night and I would miss it. Twenty-five years later I make up for it at its 50th anniversary.

Pete assigns us our jobs. He will file the flight plan and I am to oversee the refueling. Roz has taken care of the paperwork today — showing of passports and licenses. We hear that Sheila Scott has a fuel leak and may be on her way back to the Islands. She had taken off for San Francisco.

Sunday, June 12, 3:30 a.m. I am ready to go when Pete stops by. We go to an all-night restaurant for breakfast and to order box lunches. Then off to the field where the customs man is waiting. I ask Peter where to hide my fruit, but he says it is all right, leave it where it is. Roz has already hidden their fruit and mine is promptly confiscated except for the peeled pineapple. The customs official admitted it was stupid because we'd never get to Australia with it, having either eaten it or thrown it out, but he had to do it anyway.

Chapter 15

June 12, 1966. Honolulu, Hawaii, to Canton Island. I check my plane thoroughly inside and out while waiting for Peter to come back with the weather report.

I dread this part of the flight. We flew 15 hours from Oakland, California, to Honolulu, Hawaii, over water with no alternate landing fields, but the Islands are large and form a string that lay across our path.

Navigation to them is relatively easy; you might miss your exact planned spot, but given sufficient fuel and the many radio aids available in 1966, it is hard not to find them if you are a competent pilot. Although clouds covered most of the ocean below us, they were fluffy cotton balls, not the thunderclouds that lie ahead.

There is so much airline traffic going to and from the Islands, it gives one the impression there is help near, whether that is true or not.

The Honolulu to Canton Island flight will be 12½ hours over a desolate ocean. No November is stationed out there.

This time we are heading straight into the intertropical convergence at the equator and on the international dateline at 180 degrees longitude. The most monstrous storms in the world develop there.

I know that Peter and Roz won't leave Honolulu unless the forecast is for three-eighths or less of cloud cover for this convergence area as it can rapidly develop into a six-eighths cover. By the time it will take us to get to Canton Island, the weather can become exceedingly threatening. I am nervous, knowing this.

In truth, I have a terrible feeling of dread — the same dread I remember feeling in 1946 as I sat in a streetcar that crashed head-on into another streetcar in the fog and left me with a fractured vertebrae.

Roz described it to me in Oakland, "Here is the most intense heating, the greatest buildups in the world, big raindrops and big hail, heaviest downpours. There are storms all around, with no hope of going over them. Layers everywhere, and even to get through the lower parts of the "holes" you might have to be at 20,000-23,000 feet. These clouds often rise to 70,000 feet.

"Pete has radar in his plane, so we can be guided around the worst of them. We have detoured as much as 200 miles to the right of the track in going through the intertropical convergence, then, on coming back to our track, find we had to go 200 miles to the left in picking our way through."

At breakfast that morning I had made certain I understood how to use the HF and ADF to guide me to my friends if we became separated. Roz tried to reassure me that this will not happen.

I finish checking over the Aztec as Pete and Roz walk to their planes. I rip the seat of my slacks as I climb into the cockpit; is this a harbinger for the day's events? We fire up our engines. My landing gear green light does not show. I look at it closely to see if it is just faint. It is not working. Everything else is okay.

I call Pete over. He suggests the battery is low or a circuit breaker out, but I point out that the rest of the circuit is okay — those lights are on. We taxi out. On the runup my left magneto shows a big drop on the gauge, then it clears up. I sigh with relief. A fouled plug had been a concern on the flight to Honolulu and one of the items we had checked the day following our arrival.

Pete takes off first, then Roz. The air is still so I wait a few seconds longer than I had at Oakland for her plane's prop wash to dissipate, then I take off. The beast does its usual sashaying back and forth as I slowly push both throttles forward. I am happy about one thing — we are using a much longer runway than the one we had used at Oakland.

As the plane gathers speed, I desperately try to keep it from bucking. It becomes difficult to keep on the ground; I lift it off 17 miles per hour slower than I wanted to. I keep it close to the ground until I have 128 mph, then gently inch back on the control wheel to gain a little more space above the ground.

At the end of the runway I push the button to electrically retract the gear and think I feel a slight lift. Still no light; I pump the mechanical gear lever just to make sure. The plane shudders violently and I look back at the field — a purely reflex action as I know I cannot safely land with the fuel load the Aztec is carrying. The center of gravity is eight inches aft and my plane carries 550 gallons of fuel. This equals ten 55 gallon drums of fuel.

The two engines burn 20½ gallons an hour, giving me a little more than 26 hours of flight, if a tank doesn't spring a leak.

We take off on the inboard mains. After gaining some altitude, I switch tanks and burn a few minutes of fuel from the top tank because of the expansion due to the lowering of atmospheric pressure, then I switch to the rear tank.

Pete and Roz are way ahead of me as we climb to 10,000 feet, and it is much more difficult to catch up with them than during the flight from Oakland to Honolulu. After an hour I get within a half-mile of them and I can relax a little.

I eat one of the restaurant's dry sandwiches and swallow a little of the distilled water. Peter and Roz are moving away from me again. I push the throttles full forward, leave them there while slowly catching up. I slide the Aztec into the third slot of our formation.

They are having trouble reporting our position to Honolulu on the HF and they ask me to try. After many attempts to write down our position while holding the controls to keep the plane steady, I finally do it and reel out the antenna. The plane continues to swerve a little because of an uneven fuel load.

The autopilot seems not able to hold the set course. I use it as little as possible. Making it more difficult, Pete in the lead plane constantly changes his course (even though he later denies it), so I have to keep readjusting the autopilot.

I reel out a little more antenna as I call with our position. There is no answer and the output meter shows no indication of anything being sent.

Pete and Roz are still climbing, fading now into the horizon. My plane will not climb as fast as theirs. I call Pete and tell him that even with full throttle I am having difficulty keeping up with him. He says "Roger" and keeps on fading away.

I tell them to slow down. As I try to call our position once again on HF they disappear. I am now at 13,000 feet.

I am not too concerned at first because I feel that if they slow down I can catch up. I tell Peter that, with a few unladylike words added. He says they have slowed down. They may have backed off about 50 rpm.

I am using full throttle and keep the aircraft as straight and level as possible. My plane simply climbs too slowly. It is a different model than those flown by Roz and Peter.

It's great when you fly over in a commercial jet at 48,000 feet. You see the tops of the clouds at your level. When you're down there at 13,000 feet you are among everything and you are looking up at these tremendous columns, boiling, moving, changing. So impressive. You see how vast it is with that weather, the tropical front, and you are low enough to see the waves with their white tops and how strong the wind is. Impressive, yes, but frightening. I have my oxygen mask on for awhile to be sure I stay alert.

Years later in a conversation between Rosalind Merrifield and my sister Betty, she added the following comments about our flight formation: "Peter was in front, I was behind him on the left-hand side and Claire was behind me on the left-hand side. Peter could look back over his left shoulder and see us.

"This way, if we had some sort of radio malfunction and were not in contact, we could always look back and see that the person was there. If not, then we would have to start looking. This is one of the beaut parts of flying in formation — you are your own search and rescue team. Well, search, anyway.

"We had enough fuel to orbit at almost any point on the flight, not just to dead center, until the Coast Guard or some other aircraft could take over. There would be a positive fix on the person who was down, which is the hardest part. You can have equipment on your aircraft which allows the other aircraft to home in on you, but they have to be over you before they can find you. Even then, in a high sea, they still might not find you."

Roz continued, "There was a Japanese-built aircraft, a turboprop job, one that carries forty to fifty passengers. They were being ferried out three in each formation, from Japan. One fellow went down not too far out from Honolulu. The other two aircraft orbited him until one of the Coast Guard planes reached them. The CG plane did not have the specific pin-pointing equipment needed on board, but they got there in a hurry and the other two ferry aircraft could go on.

"While the two aircraft orbited the crippled one, the pilots watched the downed pilot ditch the plane successfully, saw the fellow crawl out of the aircraft, inflate his life raft and climb into it. He then established VHF radio contact with them. Then the Coast Guard came and they left for Honolulu. "The CG continued orbiting the VHF signal during the night. Then the radio went out, for whatever reason.

The CG orbited where they thought the life raft was until the CG with the accurate pin-pointing equipment arrived, about 3 or 4 a.m., and, in the morning when it was light, the downed pilot and raft had disappeared. They never found him.

"It is important to be in a diagonal formation and that we stay in sight of each other. To do that, unless you have a really good autopilot, you have to fly the plane on the step.

"Because of the fuel load, the center of gravity is way back in the aircraft and it is easy to get it off the step. The plane starts mushing, and suddenly everyone is far ahead and it is hard to catch up. It may take 45 minutes to catch up.

"Peter is a bit of an old mother hen. He wanted to look out the window, see us, and go back to his sun shots and navigating.

"We wanted to be within a half-mile of each other. Claire was to stay back and look out the right-hand front window.

"That's the ideal position, especially if you are down a little bit. Unfortunately, Peter wanted us up a little bit where he can look out and see us! If you get too high you can't look down and might overfly the others. You might put your wing down and hit somebody.

"Two aircraft are best," said Roz. "A third pilot has a hard time flying formation. If the second person gets off the step a little bit, then the third person slows up and meanwhile the first one gets too far ahead and both have to try to catch up."

My Aztec begins to roll violently to the left and I hastily bring the wing back up. It is increasingly left-wing heavy. I try to trim it while keeping my eyes searching the horizon.

I keep to the 200-degree heading Pete gave me but I am afraid that it might be seven degrees off from the heading of 207 degrees we had when I was last near them. He might have changed his heading. His heading constantly changed back and forth about ten degrees during the flight. Turbulence bounced us around, making a set heading even more difficult.

Pete began describing fracto-cumulus clouds to me. I believe I see the one he is talking about and head for it. Roz describes a little vertical cumulus on one side of a big hole; I see it and head for it.

Two planes which appear to track close to the same line may actually be far apart. A mile apart is too far to maintain visual contact. I think the two planes are to my right and far ahead. All I can see are clouds and an empty sea.

I switch to the front tanks; the rear tank now has four hours out of it. Pete radioed to me to pull the pin on my emergency transmitter. I do so but he can't get a bearing on it. I suggest I home in on his HF.

I tune my ADF in on his signal and it indicates he is behind me 180 degrees, which doesn't make sense. Roz then tunes in hers and it shows 56 degrees to my right. That is more likely.

I ask Roz to try homing in on my HF, but we find that it isn't putting out a signal. They say I am hard to understand on the radio. No wonder — my mouth is so dry.

Peter tells me to maintain between 198 degrees and 200 degrees and fly towards an opening in the clouds on the horizon. Nasty looking thunderheads appear far ahead on our path.

Then Peter calmly said those chilling words which, when the hero says it in a novel, means a catastrophe is about to happen: "Now, Claire, listen closely to what I have to say. [Oh, my God.] Maintain your heading. After seven hours have passed, tune in 371 on your ADF. That will be Canton Island. Our problem is that we are approaching the intertropical front. That could be trouble."

I reply as calmly as any fiction heroine would: "Okay, Peter," and write it down. I think of Sheila Scott and all the others who have flown the Pacific and made it and I say to myself, "I can do it." Then I remember the many others like Amelia Earhart, who, even with the highly competent navigator Fred Noonan with her, didn't make it, and I feel a stab of fear.

Those nasty towering thunderheads on the horizon are closer and I said plaintively, "Please don't leave me out here alone, Peter."

He said something I didn't quite catch, but it sounded like, "But what can I do, Claire?" Then he added, "Besides, we must hurry or the dining room will be closed at Canton Island."

The dining room! Oh, my God!

Then Roz breaks in and describes some cirrus she said they are under. I see what I think are the clouds to my right and ahead. I ask her to try the HF again. This time the needle points to 30 degrees to my right and towards the cirrus.

I tell Roz I have a good bearing and if they will orbit under the cirrus I think I can catch up. She tells Peter they are going to orbit. Orders him to orbit. A few minutes later I see his plane flying at my right, and I shout, "I see someone!"

Roz pulls up behind me and then ahead into position. I still have full throttle, but ask them to throttle back a little so I can keep up. I reel in the HF antenna and am able to keep up fairly well after that.

Then tears come, the relief is so great. I have spent an hour alone over that endless sea, something I hadn't planned to do, hadn't prepared for, when I began this flight from Oakland.

I check my wing tanks and find what I had suspected: the right outboard tank has leaked all its fuel. I had fought the badly out-of-balance condition while working with the radio and trying to catch up to Roz and Pete.

I switch the booster pumps ON, put the right wing tank in OFF position, put the left outboard main ON, put the cross-feed ON, and turned the front cabin tank OFF. Then I turned off the booster pumps one at a time. I can now burn off the left outboard tank and even the weight, putting the plane back into balance. It will be so much easier to fly!

Roz calls out our position to anyone listening, but no one acknowledges. Pete says this is the loneliest spot in the world. Amen! There is literally nothing but millions of square miles of ocean.

After an hour and a half I switch ON the boosters again, put the front tank back ON and turn the left outboard tank OFF, then the boosters OFF. The plane flies even better. Earlier, I had tuned in Canton Island on the ADF and I am continuing to receive a strong signal, indicating we are about five or six hours out.

We fly around thunderheads and over some heavy rainstorms as, traveling west-south-west, we cross the equator an hour from Canton Island.

Twelve hours out of Honolulu the small coral atoll comes into view. The great thunderheads we saw in the distance are beyond it, not in front as we had feared.

Roz flies in first, landing on Runway 9; Pete is high and he takes a little while to land. I seize the opportunity to fly over the atoll and take pictures with my 35mm still camera.

My ADF had worked perfectly and I could have found the island alone, if it had been necessary. I'm glad that it was not. How frightened I had been!

I call Pete and tell him I will lower my gear, fly over the strip and for him to take a look. Roz and he report all looks okay. Roz suggests I turn off the navigation light switch. I do and my green gear lights come on; I blush. In the Aztec, unlike the Apaches I am used to, the gear lights are not just dim, they don't light up at all with the navigation lights on. At least, this one didn't.

I land, taxi up and park neatly beside the other two. We had a nice welcome from island personnel. There are 80 men on the island, no women.

Pete comes to help me out of the plane, spots the ripped seat of my slacks, laughs and then tries to get me to come over by the wing. I say that I can't in my present condition.

He carries a water jug. Now he lifts it over my head. I don't mind the dunking. The temperature is about 90 degrees here, just a little south of the equator, and maybe nine feet above sea level. He then presents me with a weather chart properly endorsed for crossing the equator for the first time and I happily put it among my souvenirs. A handsome Gilbert Island native hands me a sweater which I tie around me school-girl fashion.

We get our baggage out of the planes and head for our rooms in a nice house which had been fixed up recently. The decor is simple and unadorned, but it is neat and clean. I change to a lighter pair of capris, then we dash off to the dining room, which is about to close. I haven't been to the little girls' room yet, but I have to wait until after dinner.

Marvin, a NASA tracking station employee, invites me for coffee at his house and then to take in the movie, "The Sergeant was a

Lady." I accept and go back to my room to change to my shorts and a brief top. It's too hot for anything else.

After the movie is over, I go home to take a shower and go to bed. I find hermit crabs scrambling all over the porch and yard and out across the road. Hundreds of them. I am amused by their antics. Each one, of course, carried its protective "house" shell that it had moved into after a previous owner abandoned it. As the crab grows, it moves from its out-grown shell to a larger one. I wonder where they had come from; I hadn't seen them earlier.

In bed, propped up by pillows, I enter a few lines in my notebook of the day's events. The gas tank leaking, causing me to fight the controls, our radio communication problems between the three of us, and my being out of visual contact with Pete and Roz for an hour, my tears of relief when I caught up with them.

Now I understand how frightened Amelia Earhart was on July 2, 1937, when she could not find Howland Island and faced her death.

About 400 miles northwest of Canton Island are two small coral atolls, Howland and Baker, 38 miles apart. The Navy built an airstrip on Howland in 1937.

Howland is approximately two miles long by a half-mile wide, with a maximum elevation of 20 feet. The Coast Guard cutter Itasca was stationed there because of Earhart's flight, to help with radio transmissions.

Twenty years after my flight in 1966, Amelia's sister, Muriel Earhart Morrissey and Carol L. Osborne, published the best and the most comprehensive book I have yet read on her life and the events leading to her death. *Amelia, My Courageous Sister,* includes all of the Itasca's records regarding the tragedy.

Amelia Earhart and her navigator, Frederick J. Noonan, left Lae, New Guinea, at 10:00 a.m. (local time) on July 2 for the 20-hour, 2500-mile flight to Howland Island. They were to refuel there, fly to Honolulu and then on to Oakland, California, to complete the first flight around the world by a woman.

They left Lae under several bad omens: they both were ill from gasoline fumes, travel-weary from problems encountered almost everywhere, and the radios had given them on-going troubles. Amelia remained fairly ignorant of how to use the radio. In the end, this killed them.

Fred Noonan was a superb navigator, having spent many years with Pan American Airlines instructing all their navigators on the transpacific run. He navigated on all the PAA pioneer flights across the Pacific Ocean.

They took off with 1100 gallons of gasoline, enough to carry them 4000 miles if they didn't run into storms or strong head winds (which they did); more than enough to reach Howland even if they had some difficulty with weather, headwinds, or navigation.

At 2:45 a.m, on July 2, (they had crossed the international date line, and since they were going west to east, they picked up a day, so the flight started on July 2 and ended on July 2), the Itasca heard Amelia's voice for the first time. "Cloudy and overcast," she said.

The weather was clear with no clouds at Howland Island and none in any direction except about 50 miles distance to the northwest where a heavy bank of clouds was reported by other aircraft.

Apparently Amelia's radio receiving equipment was not working or she was not using it on the correct frequency. Only once during the entire flight did she acknowledge receiving the Itasca's signals, and that was on a frequency good only for 30 to 35 miles distance from the transmitter.

Arriving 50 miles northwest of Howland Island after a 20-hour, 2500 mile flight is great navigation. They arrived at dawn so that Noonan could get a sun sight at the instant of sunrise. But then the sun blinded them and they didn't see the ten-mile long trail of heavy smoke the Itasca's funnels poured out. Earhart reported they were flying at 1000 feet; this is much too low to spot smoke or Howland Island.

At 7:44 a.m., Earhart reported her fuel was running low.

At 8:00 a.m., the Itasca received a message from the plane that they had received the ship's signals. These were sent on a kilocycle band good only from 30 to 35 miles. The plane circled for an hour at the same signal strength, then started flying away.

Why they continued to circle for an hour over an empty sea is one of the puzzles that may never be solved. Obviously, they knew they had arrived where they should be. But where was the island?

At 8:44 a.m., Amelia sent her last message, that she was running north and south. Itasca radio operators reported that Earhart's voice, for the first time, sounded frightened, her voice broken and choked. Then, for some reason, she again changed to a different radio frequency. That was the last message ever received from her. All of the supposed messages and signals reported later proved to be false.

The Itasca commander believed Amelia probably went down within 100 miles of Howland Island and that her plane was submerged within minutes after her last radio message.

That was in 1937, and in 1966 we still have radio trouble even though we are expert in using them and have three sets in each aircraft. And, I was frightened, too, when I found myself alone over an empty sea.

Chapter 16

Monday, June 13, 1966. Canton Island to Guadalcanal to Australia. We have an early breakfast and go out to the aircraft to check them over. We will leave at dawn tomorrow for Guadalcanal in the Solomons. By the time Pete is satisfied with the planes, it is time for lunch.

We drive out along the north side of the atoll in our borrowed car. I pick up shells and coral, some with purple stems, with Pete saying again and again, "Enjoy them now, Claire, because you can't take them with you!" He means it, too, but Roz over the years has found hiding places and I figure she will help me smuggle them to Australia.

Weight is an important item to Pete — and it is important, for each gallon of fuel equals six pounds — but he goes too far sometimes. Like the time he prevented Roz from buying a lovely Japanese glass fishing globe with the netting around it that she found in the Canton Island store on one of their ferrying flights. On three flights she looked at it, and on the fourth, when she was determined to get it anyway, it had been sold to another ferry pilot.

I am badly sunburned by the time we end our tour. While I wash my clothes and nap, Pete and Roz swim in a screened pond fed by sea water. After dinner, we drive out to the wharf where I take pictures and do some fishing with a borrowed rod. All the fish get away — they must be really big ones to get away, you know.

This atoll is part of the Phoenix Group of eight islands in Polynesia on the direct route between Hawaii and Fiji. Since 1958, commercial and military long-range aircraft fly over Canton, but in 1966 it was still important because of the emergency landing facilities and the NASA tracking station.

Sometime later, Canton Island was closed to all air traffic because airliners no longer used it as a refueling stop and Bendix's contract with NASA had ended.

The 1935 flight of Pan American's China Clipper from San Francisco to Manila via Honolulu, Midway, Wake and Guam, gave new meaning in our Western view to all the tiny islands in Polynesia. Instead of as guano, pearlshell, and copra sources, they were seen as emergency landing fields.

Canton Island has a few trees on it, planted under U.S. Government direction, in hopes they would multiply. Years later I met the man who was in charge of that program. As I hike along the beach, I think that it must take great courage to live on such an atoll.

During storms, I am told, a whole island will vibrate from pounding by heavy seas, and huge waves sweep over some of the low islands. Almost none of the atolls are more than 12 feet above sea level unless building material is brought in for a landing field or soil in which to plant trees and vegetable patches. Only coconut palms will grow in salty soil, I'm told.

As I write this in the 1990s, I think of the hurricanes on or near our own Atlantic Ocean shores which strike with such great devastation. Many of the people living on small South Seas islands do not have quite the idyllic life we like to imagine.

About 50 or 60 American and British men are on the island to aid the NASA tracking station and airfield facilities and a small group of Gilbert Island native men were brought in to help. No women; women cause trouble; children require schools.

We have ice cream in the restaurant. On my way back along the footpath, I see the hermit crabs again. They come out of the lagoon at night (one can actually hear them come out of the lagoon) and crawl across the island through the brush by the hundreds in their nightly scavenging. Biggest ones I see are three or four inches long. They go back to the lagoon before dawn...otherwise a bird or rat or other predator eats them.

Tuesday, June 14, 4:20 a.m. I'm up this early as Roz and Pete want to be called at 4:50. I usurp the bathroom until then. I packed last night and am ready to go. Pete and I go out to the aircraft. I drain the sumps on my plane, check the oil and gas, and give it a walk-around check. I stow my gear and get the cockpit as ready as possible because when they decide the moment has arrived for take off, they go.

We went back for Roz, who by this time has finished her shower. Off to a very nice breakfast at 5:30. The cook has gotten up a half hour early to accommodate us. We bolt our food, as usual. A last trip to the little girls' room and we are ready. We dash to the planes.

Pete had made out the flight plan the night before as the weather usually remains constant, either bad or worse. Two fronts lie parallel to our course; we are going down the trough between. Only scattered clouds are over Canton Island itself.

I am number three for take off and even though Pete told me he is going to go right, he goes left. Roz makes a wide turn and I shorten my turn to close the gap. Roz asks me if I can see Peter and

I can, but then she spots him right away. Several times they lose sight of each other as we climb to 8000 feet. I have both in sight all the time. I make sure of that!

We are in constant radio contact with each other and I am, this time, able to help them find each other, not an easy task as we dodge among layers of dense, dark clouds boiling up vertically and horizontally.

Flying in last position is especially difficult in this terrible weather. It is now an even looser formation which leaves me away out at the end. If the lead plane turns left, I try to anticipate and move left, then Pete moves to the right and I am suddenly out of his line of sight and in trouble with him. It's a longer distance to catch up. I can't fly close to him because then Roz won't have room to maneuver.

It's impossible to use the autopilot for this type of flying except for a few seconds at a time, because the lead plane cannot fly straight and level, and Roz's distance varies quite a bit, too, as her plane bounces around. Roz does an excellent job and certainly has this particular type of flying down well. She knows what she is doing. But despite all that, distance, altitude and direction varies considerably between us all as the turbulence bounces us and we twist and turn among the clouds.

This puts me at a distinct disadvantage. I found that for me the best spot is the slot between them and behind, then I have both in sight, but when Pete changes his heading he can't see me, and it constantly changes back and forth. A few times he told Roz he couldn't see her. The weather is terrible with monstrous clouds and heavy rain.

We cross the international dateline that sits on the 180th meridian at this point, and we go from Tuesday to Wednesday and from Polynesia to Melanesia.

There are two frontal systems parallel to our course — one on each side of us. Sometimes we're between them, sometimes working our way through the mess. Pete has Bendix Radar equipment so we are able to get around the worst spots. I take pictures whenever I can manage it.

After 11 hours of flying, we see Malaita, one of the ten large islands of the Solomons, and that tells me we are just 30 minutes from Guadalcanal.

The Solomons consist of about 300 islands with several hundred more that are barely visible above the sea. About half of the population lives on Malaita, a beautiful island that once was considered the most dangerous in the Pacific for its head-hunting rivalries and cannibalism.

Since World War II, visitors have been welcomed in a friendly way. The smiling, generous young Americans coming ashore at Guadalcanal won them over.

The rest of the islands are hidden by the greatest, blackest clouds I have ever seen. Roz and Pete wend their way through the cloud canyons, and I close in on Roz. She has flown these islands with the geophysical survey group contracted by the United Nations. I know she knows where the airport is.

We slip down over the hills, down to one thousand feet above Purvis Bay, and glide over Guadalcanal. Henderson Field is ahead on one of the northern slopes of Guadalcanal.

Roz lands first and I circle around the field taking pictures, then Peter lands and I finally decide to put away the cameras and land. I always have to be last, anyway.

The airport itself is clear but surrounded by angry storm clouds. I take more pictures of this famous field after I land. The field has a coral base with grass growing on it. It is beautiful in its setting against the hills.

We are met at the airport by the group of Swedish and German pilots employed by the geophysical survey group. The Aztecs they fly are rented from Avis of Australia, and this is the reason we went into Guadalcanal in the Solomons instead of into the Fiji Islands, a more direct route.

Pete needs to check on the organization and to see how the planes and the pilots are performing. The pilots drive us through areas still showing damage from World War II, and then to the Mendana Hotel in nearby Honiara, the capital.

The hotel is near the center of town on the beach. Today, in the 1990s, it is a first-class hotel. There is a swimming pool; all rooms are fully air-conditioned and with showers, toilets, and telephones. It has a good restaurant.

But that is now in the 1990s, and this was 1966. My room is clean, plain, and spare with just basic furniture — a bed, one chair, a small table and a washbasin. The huge fan in the ceiling of my room and the room's plain appearance remind me of movies I have seen that were set in the tropics. I can't find the women's restroom and shower, but the men's is just next door. I hope traffic is light.

A new wing is being built onto the hotel but the plumbing in each room is not completed, and the rooms are dirty with construction dust. We stay in the older section.

My room has been freshly painted. Pieces of cloth that perhaps had served as drapes were on the chair. Nothing covers the windows. Melanesians, from their village compound right outside, watch everything that goes on in my room.

Workmen had repaired the railing on the veranda in front of my room. If these carpenters are like those at home, they have dropped nails everywhere. I go out on the veranda and collect a handful. Using the butt of my hunting knife as a hammer, I nail the drapes over the windows.

The English, Swedes, and Germans working for the geophysics survey group have invited us to have dinner with them at the hotel that night. I will have to dress, as in dress and high heels. I am fortunate to have a washbasin in my room; I manage to make myself presentable.

I walk carefully in my high heel shoes on the pounded coral pathway and into the large lounge overlooking the sea. The nice young group of people greet me and offer to buy me a drink.

I suggest Bourbon. An embarrassed silence follows, a clearing of throats, and a high-pitched male voice said: "Bourbon in British country?" Bacardi? No. Tom Collins? Yes.

Melanesian native men serve our drinks and later our dinner. Bare from the waist up and barefoot, each wears the native sulu, a long strip of brightly-colored cloth wrapped skirt-like around the body and tucked in at the waist. It is similar to the sarong. One native sits on a bar stool softly singing "My Darling Clementine" to himself.

In the late 1980s, the intelligent and adaptable Melanesians got rid of the British (peacefully) and set up their own government. Maybe they have now imported Bourbon for the American tourists!

It begins to rain softly, growing steadily harder. It doesn't bother the dozen of us there for dinner and we have a good time.

I listen to the stories of the men and am reminded of the ones Roz had told me about her days of survey flying.

She had operated Aztecs in the Solomon Islands under conditions that were actually worse than crop-dusting. She had to fly on set headings, north and south, maintain a certain altitude, and with the mountain coming up, hold until the safety factor was almost gone before she made her turn.

The planes carried cameras, magnetometers, various other instruments. Two aircraft flew together, one behind the other. And while she was flying as fast as she could in this straight line, she still had to fly for the slower aircraft in order to maintain the right distance between them.

It was pretty frightening for the second aircraft pilot because when he saw a big hill coming up he must start reducing power as he reached the base of the hill. He must also anticipate the deceleration of the first aircraft so as to maintain the right distance.

It is a terrible feeling to reduce power while approaching a huge hill. Both pilots have to go up as close as they can, do a breakaway, build up speed, go into the turn, then pick up the line again as they fly back.

The pilots have to make as few breakaway turns as possible. The company is paid only for the approved line miles flown so all excessive turns cuts into the profit.

After flying under those conditions, Roz knew that the Aztecs could really take it. She'd operated them full throttle all the time, full rpm, just one prop pulled back a little bit to synchronize the two engines. Normal operation is at 2400 rpm and they are flown at that on ferry flights.

The survey group was trying to find more mineral deposits. In 1955 gold was found just twenty-two miles from Honiara, and undoubtedly other minerals are there in the hills.

After dinner we stroll out of the front entrance of the hotel. Just beyond the lawn is the ocean. A swimming area had once been fenced off, but much of it is broken. It is a shark fence; I am told that the sharks in the Solomons are quite mean. Along the coasts of New Zealand and Australia there are many places fenced off for swimming. I do not want to think about sharks.

Roz and I discuss our strategy for the next morning. As I have said, the men's shower and toilet is next door to my room, right at the end of the veranda. We've learned that the women's room is about 100 yards from us and well on the other side of the hotel. We plan to get up early, make our dashes to the men's room, and get out before being caught.

I'm to go early, about 5:00 a.m., and as soon as I leave, Roz will go in. This works pretty well, but there is one little Englishman who probably hated women in general anyway, and who now hates women pilots in particular. Roz catches him in the shower room the next afternoon and I catch him in the etcetera.

Roz told me of the time she, Betty Faux, and Peter had stopped in the Fiji Islands on their flight to Australia. Betty and Roz were shooting the tropical scenery around the hotel, the palm trees, vines, and flowers, while Peter was taking a shower.

Roz was rewinding but Betty was shooting when Peter came outside. He has orange hair and white skin, and wrapped around him was a large pink towel. With a big grin on his face, he stood tall, threw out his arms and said, "Here I am, girls!" His towel started to slip and this huge man shriveled into an embarrassed heap.

Much later, trying to record the day's events while sitting up in bed, I fall into a deep sleep. The pen and paper are still in my hand and I am still sitting up when I wake about 4:00 a.m. The day's events and 11½ hours of flying had finished me. I get up about 5:00 a.m., as planned, to visit the men's room. I breakfast alone and then, as if in answer to my prayers, one of the men's wives, a German girl, drives up in her little car and offers to take me sightseeing. We spend the day filming market places, bush country, war relics, and gorgeous tropical rivers and coconut groves.

Market places are always interesting to me and I try to visit them in foreign countries whenever I can. I like to see the wonderful fruits and vegetables they grow and forage for, and the things they make to trade with one another.

The British had built a roofed-over area with a concrete floor on which the natives could put their trade items. The Chinese have their own little enclosed shops elsewhere, but the native women prefer to sit together as a group so they can talk. Market-day is the only time most see each other.

As we walk through the market I notice that the natives have a strange coloring to their hair. I ask if it might be one of their religious or tribal customs. Not so. I am told that they have been watching movies and had put peroxide on their black hair. Instead of blond, it turned a reddish tint. I even saw babies with this coloring.

The German woman drives me down to see the wreckage from the war. After that, we drive through some extensive coconut plantations. Near coastal areas the untouched jungle is quite thick, while in the hills it is more open. Ravines are filled with trees; native huts are scattered among them.

We go on towards a place where later in the afternoon we are to meet Roz and Peter. We stop to look at a beached military ship, then we come to a place where water flows down a mountain ravine, through a large pool, and into the ocean. This is a popular swimming area and it is gorgeous, just like the photos in the travel magazines.

Roz and Pete arrive. The three go swimming while I watch them and take movies. It was one of the most glorious places I have ever seen, exquisite with a backdrop of palms, vines, and massive flowering shrubs.

Guadalcanal is a continental island, not a coral atoll like Canton Island. Continental islands are the exposed peaks and ridges of mountain ranges crossing the floor of the Pacific Ocean and are kept green by moisture-laden tradewinds. If you're checking a map, look along the 160-degree longitude line below the equator.

I have added to my Canton Island sunburn and blisters now cover my upper arms. After dinner that evening, Pete, Roz and I plot our course to Townsville, Australia, across the Solomons and the Coral Sea. Pete purposely does this in full view of the other diners — he likes to show off a bit. We have spent a full night and a full day in Guadalcanal. We are to take off tomorrow morning for Australia.

I have planned an early-to-bed for me, but as usual, it is late before I have my clothes packed and am in bed. Four to five hours of sleep a night is all I have managed so far. However, this isn't too far out of the norm for me.

June 17, five a.m. It is very humid this morning. I have a hard time keeping the windshield clear enough to see to taxi because of the moisture gathering on it. All systems go, we take off. Roz waits until Pete clears the end of the runway, then I wait at the start of the runway until Roz is clear.

Cumulus clouds are building rapidly around the islands again. We circle over Purvis Bay, climbing until we reach 13,000 feet before we head out across the Coral Sea. The air is turbulent. We are on oxygen.

Roz had explained it to me in Oakland: "Any altitude over 10,000 feet means caution to a pilot. Oxygen deprivation is serious. The Aztec flies full-throttle at 8,000 feet, so we try to stay between 8,000-10,000 for oxygen requirement and for the speed.

"On a previous flight with Peter the clouds were slowly rising and I was just following Peter. This went on for three or four hours while we slowly climbed. All of a sudden I feel that something is wrong. The engines? No, the engines are perfect. I grabbed the oxygen and started breathing it and I was all right again. But all the warning I had had was just a feeling that something was wrong.

"Some years ago Peter took up two parachute jumpers who wanted to make an altitude record jump from 22,500 feet. On the way up, instead of putting their oxygen on, they wanted to save it. Peter realized his own oxygen was not working, looked back at the jumpers and saw that one of them had slumped forward, caught his parachute in the oxygen lead line and had doubled it over, closing it off. The two blokes had passed out.

"Peter put the plane into a dive. At about 17,500 feet they started to come out of it a little and asked to be taken up again. Peter refused. So they said they'd jump out right then, and they did. Peter couldn't prevent them because the back door was off."

We head directly for Townsville. Navigation is not difficult — Australia is a very large place, but I am again having radio trouble. Every time I try to transmit, power cuts out and I can neither hear myself transmit nor can I receive. This problem becomes critical as we approach within 40 miles of the Australian coast.

The Australians consider a flight like ours to be IFR (conducted under instrument flight rules instead of visual) and we have to be constantly answering them. They are nasty about this and one must talk to them. Well, I cannot. I try both microphones and both earphones, everything, nothing works.

When we land in Townsville, their FAA wants to talk with me. Peter intercedes and explains my radio problem. I'm not actually worried about it as it is a fault in the radio and not because I didn't want to talk to them.

Then we flew to Mackay where we spent the night. There I eat my first raw oysters. After that I have oysters frequently in Australia, even buying them from street vendors.

From Mackay we fly towards Sydney. We will land next at Coolangatta for lunch. Seems they have a terrific smorgasbord.

Peter called the control tower for landing instructions, saying "This is Avis One, Two, and Three."

The tower said, "I don't know, we've already had three airplanes land here today!" Should Pete have replied, "Right, make that Four, Five, and Six"?

Again I have radio trouble. We land at Coolangatta, rush into town for lunch, then rush back out to the airport.

We arrive at Sydney airport just at sundown. I still have radio trouble and can't answer their clearance to land. I land anyway, then find out that the tower had cleared all three of us to land, a "first time" for the tower, and that I wasn't required to answer.

Peter lands first and Rosalind, second. One of the flight rules in Australia at that time was that only one plane at a time could be on the runway. While waiting for Roz to land and taxi off the runway I go around the pattern again.

There were many people out to meet us at Sydney. There was one family that had met every transpacific pilot that had come into Sydney, and they had gifts for us. The man who owns Avis, Eric McElrey, and his friend Di, are there. Peter's wife is there, and so are some women from The Ninety-Nines and from the Women's Aviation Auxiliary Organization. Roz, born in Sydney, is home.

I spend ten days at Eric's and Di's home. It is a 117-year-old house, a gorgeous place, overlooking a beautiful bay and Sydney Harbor, one of the most beautiful harbors in the world.

It has about 1000 back bays that are wooded and sheltered. Wonderful picnic areas. We spend happy hours exploring the bays in their boat.

I have my own room. It is large. The first morning Di brings breakfast to me in bed. The next morning I get up early enough to have breakfast with her in the kitchen.

Di is a wonderful, beautiful person, extremely friendly, and is a fine companion on some of my sight-seeing tours. In fact, we visit the Sydney Zoo together so I can see a Koala Bear and a kangaroo. I get to hold the Koala Bear.

One of my former students, Michael Rey, reads in the paper that I have arrived. He phones Avis and we spend a day in the Blue Mountains. We lunch in a restaurant in a little town there. There is a party back in Sydney that night, so we race back to town.

Too soon it is time to board a commercial airliner for the long flight home to Los Angeles, California. I will go back again someday to visit all the places I missed this time, I promise myself.

Years later, in 1990, I took my daughter Susan with me to Honolulu on Hawaiian Airlines. And again, in 1992, along with my teenage granddaughters Michelle and Nichole.

As we approached the Islands, I looked down and in my mind's eye I could see three small planes flying in formation just above the clouds. Tears came as I thought back to those days so long ago and my greatest adventure.

But, I have to tell you, it was a good feeling to be able to walk down the aisle to the restroom and to be served a hot meal.

Since then, both granddaughters have married and I have three adorable great-granddaughters, Mikala, Kayla, and Morgan Marie.

Chapter 17

Around 1967, the FAA decided that each flight school had to develop curriculums in the form of lesson plans for the multi-engine, the instrument instructor course, and the airline transport rating course. The other courses, private, commercial, instrument, and instructor courses, were not in the new lesson plan form, but we weren't asked to format them.

Previously, we had a simple and direct approach: In Lesson One in multi-engine, we'd list the items that the instructor would teach the student; in Lesson Two, list those items, and so on, which is very simple, practical and logical. Now, however, the FAA decided that each lesson must follow a lesson plan with the objective to be achieved in the lesson, the lesson itself, and then the performance standards that the student was to reach during that particular lesson.

Up to this time, the FAA did not have an approved school for the multi-engine rating; it was not needed. But if we wanted to be approved to teach these courses under the Veterans Administration plan, we'd have to develop the curriculum for each. Each school was to develop its own and submit it to the FAA for approval.

All flight schools who wanted to teach multi-engine flight to VA students began to build this curriculum. It was at the insistence of the FAA that the first one was the multi-engine.

After we found out how to develop the lesson and the curriculum for this course, we were to develop the instrument instructor course. Finally, the really brave ones could tackle the airline transport rating curriculum.

Well, I struggled and I struggled. I took our original curriculum, expanded the lessons, listed them out, put in an objective, and took it down to the local FAA office.

They looked at it and said, "Gee, this is just great, Claire." They sent it to the area office which bounced it back.

The area office said I wasn't getting everything in there and listing it the way they wanted it. They said I still needed to be more specific on the objectives, more specific on the performance standards, and so on.

I struggled with it. I worked all night many times because during the daytime it was almost impossible to think productively, type the material, and run a business. Either I'd write at night at home, getting the thoughts formulated, then come out to the field early in the morning to type, or, once I had everything pretty well lined up, I'd spend the night in my office with my electric typewriter.

Each time I got it into a form that looked acceptable I'd take it to the local FAA office. They thought it was great. They matched up all the maneuvers against the chart which told them how many maneuvers there were to be, how many times a particular maneuver was to be performed in the course, so forth.

I came out even with the local FAA score card. Everything was in there, but it kept bouncing back from the regional FAA office.

Semantics had reared its ugly head. A word that meant the same thing to the majority of the population apparently meant something else to the regional FAA.

A change in the word "shall" to "will" and "should" to "would", this sort of thing, was causing a problem. This delayed getting approval of the program and as long as there was delay, we couldn't be approved by the State to fly VA students for their multi-engine rating.

Time was important. I had to keep struggling and working, working all night. So, I'd retype and retype and finally get it into a form that the local office thought that the regional office would accept. Each time, it had to go to the local office, then to the regional office.

One of the local office officials commented that the regional office didn't know what it really wanted, but thought they'd know it if they ever saw it.

I was in tears half the time. I'd roar up to the local office and grimly stamp up the walk. William Glenn, manager, would see me coming, and call all the FAA men in the building to his office to hear what I was going to yell about this time.

I raged that I was doing their job for them, that they should have done this 20 years before; there was no reason in the world why I and all the other flight schools should be writing these individually. One curriculum for all would do just fine, and it should be written by one of the professors writing those ridiculous brochures sold by the Government Printing Office.

The FAA men would try to calm me down and they'd say, "Now, now, Claire, you are pioneering this."

"Yeah, well, old pioneer Claire is going to try to print these and sell them to other schools, because I don't think other schools are going to want to bother with this crap."

"Now, now, Claire, that's...that's not...you know, after all, somebody has to do this."

"Yeah, well, I'm gonna make a buck out of it as long as I have to do it."

Finally, they'd say, "Now, Claire, you are not supposed to sell them. Each school is supposed to develop its curriculum according to the personality of its chief pilot, kind of airplanes they fly, and the kind of country they live in."

I said, "You are out of your minds. You can develop a generalized test guide at your main office in Oklahoma City that has nothing to do with the kind of airplanes, or where you fly, or anything else, and it's a general summary of the maneuvers to be performed.

"The curriculum doesn't have to be developed to state the kind of airplane. It can be general enough for any airplane. You simply refer to the operation manual for that airplane.

"What do you mean, each curriculum has to be developed by the chief pilot of each school? What's that got to do with it?"

"We want each chief pilot to know what's going on in his school. If he has to write a curriculum he will know how everything is being planned."

I said, "In the first place, there are a lot of chief pilots who don't write too well and who do not understand the program too well. The job of development and writing it will be done by some public relations person or the office secretary, or some college student who may have some writing experience. You will not succeed in whatever it is you are attempting to do.

"I'm not a writer either, and this is hell on wheels to do. I think other people are having as much trouble as I am with this, and the others are going to get discouraged and just buy them instead of spending hundreds of hours as I have."

"Now, Claire, you're not supposed to sell them."

"Show me the rule, show me the law that says I can't sell them." Of course, there wasn't any.

While I was going through all this trauma with the multi-engine, I was developing the instrument instructor and the airline transport curriculums. At least, now, I had some idea of the obstacles that lay ahead for each.

Finally I got the multi-engine into the form I and the local FAA thought was acceptable. I took it to be retyped by an expert typist, then to another office with a Xerox copier. I delivered three sets to the local FAA to be mailed to their regional office with a cover letter.

This time one copy came back stamped "approved." Wheee! I was in, right? Well, no. The multi-engine curriculum still had to be formally approved by the California Educational Department. However, the FAA had told me that the State would accept whatever the regional office had approved.

I must send the original and three printed copies to the state, according to their instructions. Not typed originals, not carbon copies, but printed copies just as though I had printed them for the students. I used a Xerox photocopier for this; it made the copies look like printed ones. I don't know how we could have accomplished this if the Xerox copier had not been invented.

I sent the copies in with the cover letter from the FAA stating that my curriculum had been approved for the VA multi-engine course.

Meanwhile, Margaret Ross Berry, an old friend of mine, told me that she had access to an IBM Executive typewriter. This has a type that looks like printing. Each letter is proportional: that is, a, b, c, and most of the other letters, take the same amount of space, but "l" and "i" take only one-half space, while "w" and "m" take one-and-a half. It was really neat.

Margaret offered to type my curriculum for me, so we worked on it together. We made originals that were printing masters — neat and spaced properly, each page ready to print.

I worked at spacing on some of it the way I felt it should be; she completed the job. A local print shop made 300 copies of each page.

I had written 14 lessons for a total of 20 hours. Some pilots can get a multi-engine rating in five hours, some in ten, some take 15, but if a man couldn't quite cut it and had the money coming through the VA program, we could give him what he needed to complete his program. I had written in 12 hours of ground school and this was also approved by the FAA. The State had told the FAA early on that they'd approve whatever the FAA approved.

While that curriculum was in Sacramento, I'd been writing the instrument instructor for our school and it hadn't been too hard once I learned the system in writing the multi-engine curriculum.

It took quite a bit of thinking to get in the objectives, the lesson plans, and performance standards acceptable to the FAA. I got it typed, photocopied, and handed it to the FAA. Again it took time bouncing back and forth with semantics changes and so on, but not quite so much time as with the multi-engine.

I got a letter of commendation from the FAA regional office saying it was the best curriculum they had seen. Of course, I had no idea how many instrument instructor curriculums they had seen.

Again, the instrument instructor curriculum consisted of 14 lessons and 20 hours. Some of the lessons were for an hour and 15 minutes, while others were an hour and a half. It also had ground school subjects in connection with flight, and these I had written into lesson plan form.

The State decided that they wouldn't pay for 20 hours of instruction for either the multi-engine or instrument instruction. These two forms had been printed and I had sold copies to several schools, including some of the larger schools here in the West.

California was holding back its final approvals on the two and stated it wouldn't pay for more than either 14 hours of lessons and one hour for flight check or 13 hours of lessons and two hours for flight check on either program. I was furious.

They had not made that statement before I had printed everything and, in fact, they had made the statement that they would approve anything the FAA had approved.

I questioned them about it. The illogical answer I received was, "We have a school, Rose Aviation, who advertises that they can do better than, let's say, you can, because they can do a course in less hours, so we don't want to give a school the advantage of advertising this way over another school."

I said, "You people have to be out of your minds. What difference does that make? An approved curriculum for a course to develop a competent pilot has nothing to do with differences in prices from one school to another. That has to do with business competition, type of aircraft used, and so forth."

I said that some people need more hours than others and they should have that advantage. You don't have any buffer zone, a minimum and a maximum any more in courses, just a maximum.

They refused to budge.

I wrote to those in California who had already bought my curriculums and suggested they simply change the number of hours in each course to 14 lessons and 14 hours, and send in a letter to the state to add one hour for a flight check which they didn't have to have on the curriculum. This messed up the printing a bit.

I didn't change them on most because a lot of my courses were being bought by schools in other states such as Massachusetts and New York, as well as Puerto Rico. It interested me to find that FAA officials in different states have different ideas.

In some states, all lessons were allowed and all 20 hours; other states, ten hours. The ATR curriculum in New York had been written for 32 hours, which was maximum in California, plus 63 hours of ground school.

However, for the VA program, the State of New York said that 15 hours of flight instruction was all that would be allowed. It believed that a man who had reached the point of being able to take an airline transport rating course should know all there is to know and he would be allowed to have only six hours of ground school paid for.

I started work on the airline transport rating. That one was a real dilly because there was no test guide for the rating; none had ever been written. The performance standards were in each FAA man's head. He performed the flight test as he thought it should be. There was a considerable difference of opinion on the number of hours the curriculum should have.

I was writing the curriculum when the changes came through on the number of hours allowed on instrument and multi-engine ratings, and as I said, I was furious about that. I ran into the FAA office and demanded a meeting with the FAA men. Poor William Glenn called in all the FAA men in the building that day. They faced me once again, all grinning.

"Here I am in the middle of writing another curriculum and the State hasn't yet decided how many hours are going to be on it. I get through the whole thing, and get it all done, then they say it is too many hours or not enough hours and it bounces back; it just delays everything.

"This is ridiculous. If they can't decide ahead of time how many hours can be allowed on a course and so on, or if they are going to demand changes later after it all is done, it's all just a lot of wasted work."

The FAA men agreed and took it upon themselves to demand from California a reading and an immediate decision while I continued to develop the various subjects and lessons.

Finding the ground school subjects took a little time. All the material was not available to those of us who were not a part of the airlines and when I tried to get copies of curriculums from the airlines, I found them closely guarded.

During all this, I still gave my speeches and attended aviation meetings. I traveled to San Diego for some of this and ran into Barbara and John Tucker, owners of the Gillespie Airport Flight School. John had subject matter for all the curriculums, but he didn't know how to get them into the lesson plan forms.

He was in another FAA region and the various regions were very...didn't know what they wanted. They acted under the same theory that we had run into in Los Angeles...they thought they'd know it if they saw it.

I showed John a copy of my multi-engine curriculum and traded it for the ground school subject matter for the airline transport rating.

Finally, California said that they'd allow 32 hours maximum for ATR lessons, three hours for the flight check, and sixty-three hours of ground school. I got the form done, turned it in to the FAA, and, of course, it bounced back. It seems a word on page ten, line five, was to be changed from "shall" to "will. We were into semantics again and there were about ten or 11 errors or additions they wanted made.

I'd make them, then it bounced back again and it would say they wanted me to increase the performance standards on Lesson Seventeen or something like that. I would increase the performance standards and send it back. It would come back to me to correct something on such and such a page...and on and on.

I was getting more furious by the minute. Why didn't they catch all these things the first time? I accused the FAA of delaying action, knowing that I had these curriculums sold. I'd send in a form and the local office would say, "You know it's approved, Claire; we'll just make out the letter for you."

I had already mailed my advertising to flight schools about my three curriculums and I had many orders for the ATR. Although I valued each ATR copy at one thousand dollars, I knew most schools would not be able to pay that, so I offered it at the same price as the other two: one hundred for the first copy and three dollars for each additional copy, just about the cost of copying. I had already sold one thousand dollars worth of the multi-engine.

The FAA was simply delaying releasing the ATR since they knew that I was selling them. They also found out that I was selling them out of the region, so they wanted it to be really good. But they were delaying it, a way of slapping my hands.

I don't think they ran into too many people like me because I was going to buck them right down the line. I knew that I had a right to sell the curriculums and I was going to; their reasons for me not selling them were illogical and impractical.

They also knew that the schools were going to have to get approved in order to teach the courses under the VA program, and they were going to have to get the curriculums somewhere.

If other people had as much trouble as I did getting the material for an airline transport rating course, then some schools would never be able to get approved for it. So long as I had to do this and work at it so hard, I thought I might as well make some of my money back.

When I had completed the multi-engine curriculum and had it printed and was trying to sell it, I didn't sell it to many schools. As I said, I only made about $1000 on the multi-engine curriculum. I thought, gee, the FAA is right. Everybody else is going to write their own curriculums. Then I thought, no, maybe most of the schools will write their multi-engine curriculum because the FAA has asked them to do that one first, but when they run into as many problems as I have with the FAA and they get as frustrated as I have been, they're not going to tackle anything else and they are going to buy mine.

And, sure enough, that is how it worked. Many schools did write the multi-engine and some did get them approved; but then schools became frustrated and I sold twice as many instrument instructor curriculums. These were mailed to flight schools all over the country.

Each region in each part of the country had to do its own approving. A school didn't buy an automatically approved curriculum. They had to get it approved from their local office. To make it their own, they took off my school cover and stapled theirs on. See how simple things can be?

The FAA was holding back on final approval of the ATR. It was beginning to get a little touchy. In fact, I would send out approved sheets of partial curriculums to these people with apologetic letters and notes and explain to them the problems I was having with the FAA. I did that with as much humor as possible so that I would keep them happy. I'd get telephone calls: "Where's my curriculum?" I was unhappy and frustrated with the FAA.

Each time the new pages came back from the FAA saying, review this line or review that, I'd have to have Margaret retype that page and reprint enough copies to mail out to those holding my ATR partial curriculums. When the project was completed, I had reams of rejected sheets to cut up for scratch paper.

Well, at long last, it was approved, half approved. And I completed the rest of the printed copies and shipped them off to people. If I had already shipped off approved pages that had then become unapproved because of some word I must change, I'd ship new sheets to those people and tell them which pages to replace.

Finally, it was all done and I had written 16 hours of single engine, 16 hours of multi-engine and 63 hours of ground school of various subjects. The reason I used both single and multi is that I owned the Cessna 172 which I had fixed up for the ATR course, or ATPR course, and I did not own a twin-engine which I would have to lease.

To me, it was smart to train the student for his instrument in the single engine, making the course cheaper for him, and then complete the training in the multi-engine, bringing in the same instrument approaches, but adding the multi-engine features.

He could actually get both single and multi-engine ATR ratings if he really wanted to or just the multi-engine rating. Some men do get a single engine so that they can keep their 'writtens' in force forever; when they had the money, they'd get a multi-engine rating.

At any rate, I went over big. I sold curriculums all over the United States and to Puerto Rico. The only region where I wasn't automatically, or almost automatically approved, was the Central Region. I had no luck selling my curriculums to the Central Region people, because each local FAA office had to do the approving and they just weren't going to approve mine.

They found all kinds of things wrong with the multi-engine instrument and the instrument instructor, which I thought was pretty funny since the Los Angeles regional office had given me a letter of commendation for the instrument instructor.

I understand it is a pretty common attitude in that general region. Later, I found they had their own and was not about to let anyone else's in.

Eventually, the local FAA began recommending my curriculums to schools. They had been deeply upset with me for printing these for sale to schools. But soon they tired of reading manuscripts in all kinds of strange forms, in pencil, penciled ideas on margins and so on; not written so that they were actually readable.

They were so happy to have one around that schools could simply buy, one they had already approved, and that could be automatically sent through the works, that they began to push my curriculums for me.

I'd receive a letter from someone saying that they checked with the local FAA who told them, "Go buy Claire's; it is neat; it is done; it is great; just buy it."

I sold about $8000 worth of ATR curriculums. I did well on that. I received letters of commendation from the schools and telephone calls requesting more copies. Flight schools all over the country and Puerto Rico, as well as all the local schools, bought copies. It was very successful.

I advertised by direct mailings. A letter was printed and sent to each approved school in the United States. Individuals asked to buy — until they found the price was $100. I gave copies to close friends I thought would enjoy having them.

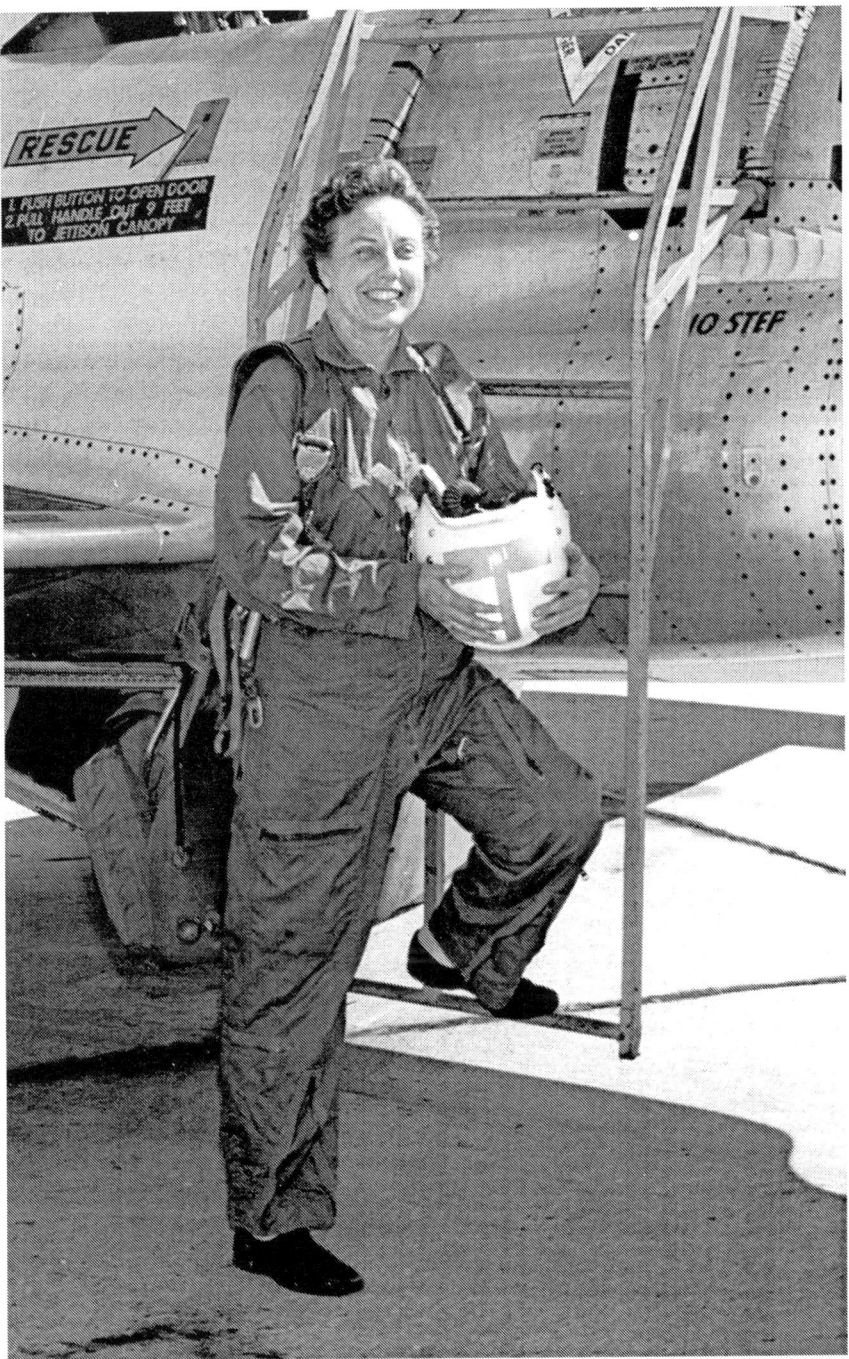

Claire is ready to ride in a T-33 at the Air Force Academy in Montgomery Field, Alabama. Her school had ROTC contracts with the Air Force, Army, Navy, and the Marines from 1964 to 1974.

New Boundaries

There's new glory in the message of the dawn
When you're flying there alone above the earth;
Then the sun comes slowly creeping
While the land below is sleeping
And a new and lovely day is given birth.

There's new beauty in the twilight as you fly
Down the lonely western skies into the night;
Then the sun's last faint surrender
Leaves the hills in dusky splendor
And the little rivers point your course in light.

There's new peace up there at night among the stars,
While the town below as lesser starlight shines.
Every pilot learns to love it
Flying high and fast above it—
And it's joy unending to the Ninety-Nines.

There's new meaning in the mystery of life—
And—fair or foul—you take it as it comes—
There's new music in the singing
Of the winds as you go winging,
And new music in your engine as it hums.

There's new magic, then—wherever we go flying,
Over mountains, deserts, rivers, palms and pines...
There's new light upon the faces
Of the gals who fly the races,
For they're bringing new acclaim to Ninety-Nines.

~ Beatrice Edgerly MacPherson

("Bea" was a 99 and an air racer.
She flew to New Horizons in 1973.)

Chapter 18

Air racing is a lot of fun. Most of the races are short one-day events, with rules set by whoever is running it. Several are long-distance, either across the United States, or from England to Australia, say, or around the world.

I've raced in ten Powder Puff Derbies and in two Air Race Classics — both transcontinental races — as well as the Angel Derby, the Pacific Air Race, and in many of our Southern California races such as the Valley Air Derby, Shirts 'N Skirts, and Back to Basics. Some are for women only, others include men.

Since 1970 I have chaired my own race for women, the Palms to Pines. I've flown in them all. In 1978 I formed The Ninety-Nines Palms Chapter in order to be sure the race would be run each year and to be able to make arrangements and solve problems well in advance of the next race.

Early in 1970, a former student, John Koich of Independence, Oregon, phoned to ask me to run a women's air race from Santa Monica to his town.

I asked three fellow members of The Ninety-Nines, Thon Griffith, Margaret Berry and Frosty White, to go with me to Independence.

We landed on a grass field. A few light aircraft were there, but no hangers, and in the town of four thousand, no hotel.

John met us at the field, and on the short ride into town, he told us that the town had purchased the field several years before and gave it to the state of Oregon in anticipation that an all-season airport would be developed. But nothing happened. He thought that publicity generated by the race would focus attention on Independence and bring action from the State.

We asked John about a hotel to house the race pilots. He said that the committee planned to ask people in the town to host race teams.

John visualized the race as a simple one: take off from Santa Monica, fly as fast as possible, and the first one across the line wins. Independence would provide food and trophies.

We told him no. Women's races had to have rules, handicaps, timers at refueling stops, and hotels for the overnight stop at

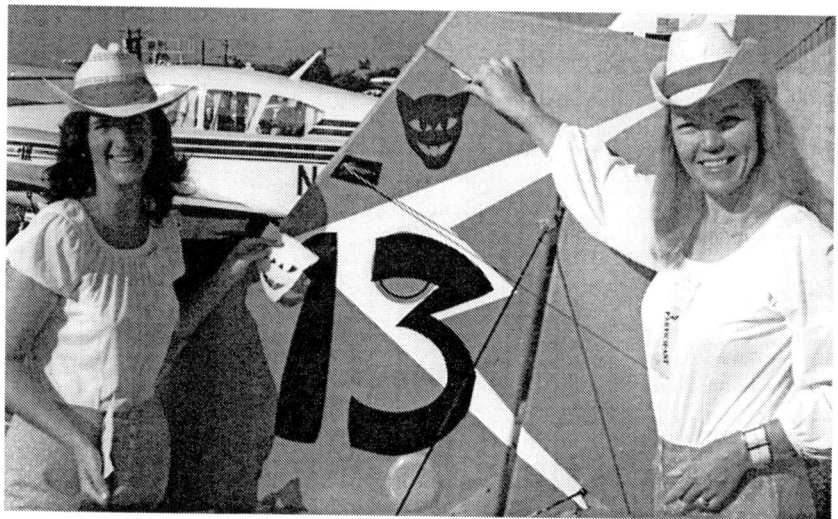

Jean Schiffmann (right), pilot, and Karen Sherman (left), co-pilot, with their entry in the second Palms to Pines Air Race. Their "Flight of Number 13" is told in this chapter.

Photo by Charles "Pat" Schiffmann. © 1999 Jean Schiffmann

mid-point for a race this long. Women would be flying family aircraft (Cessnas, Pipers, Beechcrafts, and Grummans).

John's reason for the race was to build an airport; the women's reason to race would be to have a fun weekend with friends while practicing navigation, meteorology, and so forth.

We flew home and started work on the project. The race committee set up scorers, timers, and handicappers, with race stops handled by Chapters of The Ninety-Nines nearest the route. Willamette Valley and Columbia Cascade were among the first to help.

I became chairman. I asked Margaret Berry and Joyce Failing to put together rules and regulations, using the Powder Puff Derby rules as a base. Joyce also agreed to be Chief Timer, a position she held until she moved to the beautiful Monterey Bay, California, area.

Thirty-four planes were entered and the first race was held just four months later. Margaret Mead won. Margaret was a veteran race pilot of four Powder Puff Derbies up to that time, including the 1968 race which she won with co-pilot Billy Herrin. She won our first race, and a few weeks later she again won the PPD. Movie actress and pilot Susan Oliver was her co-pilot that time. Susan was also our first Starter for our Palms to Pines.

Margaret raced again in the 1972 PPD, then placed first in the Angel Derby. She has raced in the Palms to Pines at least twice more. She is in aircraft sales, including business jets, in the Rocky

Mountain states, Pacific Northwest, and California coast, logging many thousands of flying hours.

In 1967, Susan Oliver flew a single-engine plane from New York City to Denmark. From there she was to fly to Russia, but that country had taken back its permission to enter by the time Susan arrived in Denmark, and she returned to America. She wrote about this adventure in *Odyssey, a Daring Transatlantic Journey.* In 1983, Susan was again our flag-waving starter, waving off 60 planes.

For 12 years the race went from Santa Monica to the little town in the Willamette Valley, even though the flying weather was often marginal in the valley and surrounding area.

The town was enthusiastic and quite helpful in all aspects in handling the race terminus. The host families came to the airport when the racers were due and took their teams home.

John Koich and his committee invited Oregon state aviation officials to each race event, and airport improvements were soon budgeted. The state constructed taxiways, runways, and lights. The race objective had been met.

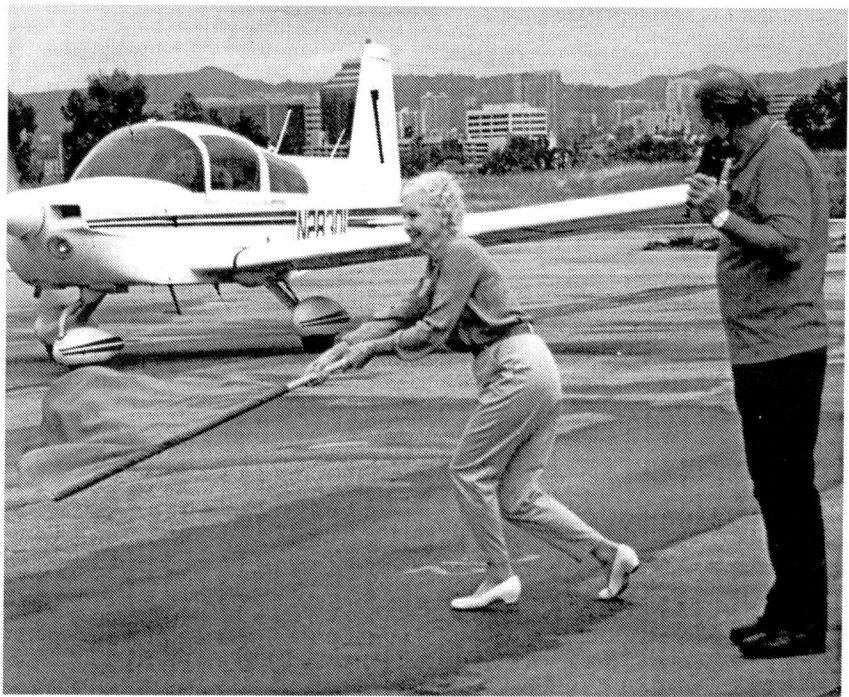

Ann Savage D'Armand, famous actress and member of our 99s Palms Chapter, flagged off the planes in four air races. Conrad Camden (right), was radio contact between pilots and tower for 20 races. Here Ann and Conrad start the 1993 Palms to Pines Race.

Photo by Betty Loufek

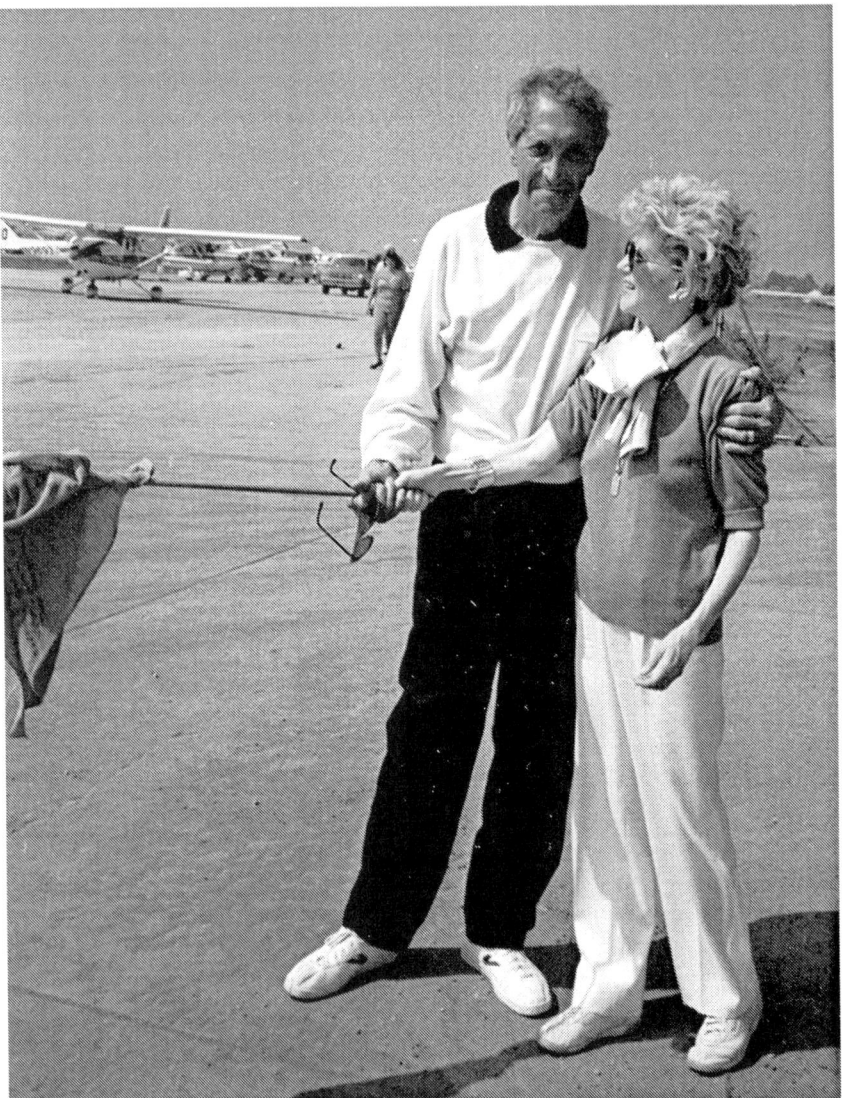

Steve and Judith Krantz practice the flag-off technique. Judith, as Honorary Starter, flagged off 48 planes. Not an easy chore.

Photo by Betty Loufek

Marginal weather around Independence had often held racers at Klamath Falls, Oregon, or at another timing point, until it cleared, but it was during the thirteenth race that heavy weather blocked flying beyond Klamath Falls for the rest of the day.

My race committee, Nina Laughbaum and Lillian Camden, and my co-pilot Rikki Barton and I were among those stranded there. We

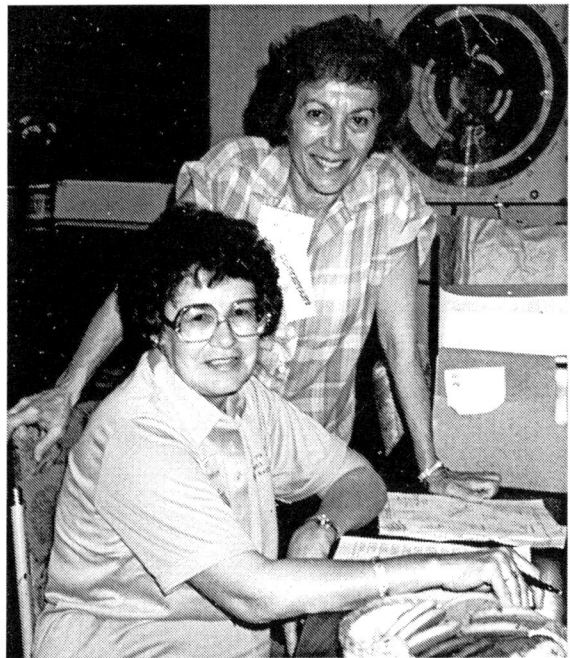

Nina Laughbaum and Lillian Camden worked with Claire Walters on her committee, then took to the air to race.

Photo by Betty Loufek

declared the race ended, and that the Awards Breakfast would be held in Klamath Falls the following morning.

While we awaited transportation, the airport weatherman Dave Williams happily handed out his wife Diane's cookies. Each year she baked cookies in anticipation of some of the race pilots stopping by, either going or coming. When the women called in on their way back home, Dave often asked about how it went, and the results. Dave is now retired, and they still live in Klamath Falls. They tell me they fondly recall those air race days.

The Awards Breakfast was held at the Copper Kitchen, who, happily for us, cooperated as there wasn't any meeting hall available. It was just great.

The gals had such a good time being all together in one place, they asked the race committee to do it again, and to find another terminus with better weather.

We chose Sunriver for the 1983 terminus; the Air Race Classic had used it for their 1982 race start. But Sunriver lies in a valley and morning fog prevented the race planes from coming in as early as planned from Red Bluff.

On arrival, the racers found themselves mingling with a flock of aircraft taking off after a weekend convention. The airport manager was powerless to stop the take offs.

Fifteen miles north of Sunriver on a dry, high plateau in Central Oregon sits Bend, the town where our parents worked for the railroad in the early 1900s.

My sister Betty, who had already scouted the area around Bend (she didn't like what had happened at Sunriver), talked the committee into looking at Bend. She also found The Riverhouse, a motel beautifully-situated on the banks of the Deschutes River.

We asked Bend about bringing the air race finish to the airport and the city embraced the idea. The airport operator agreed to close the airport to other traffic for the two hours needed, and pledged all other necessary support.

The operator long had wanted to improve taxiways and tie-down area. Now he would do it because of our race.

Donna Vasko, pilot and member of The 99s Central Oregon Chapter lives in Bend. She agreed to help with the terminus publicity and arrangements, and her fellow members would help with timing and other chores, as they had done at Sunriver.

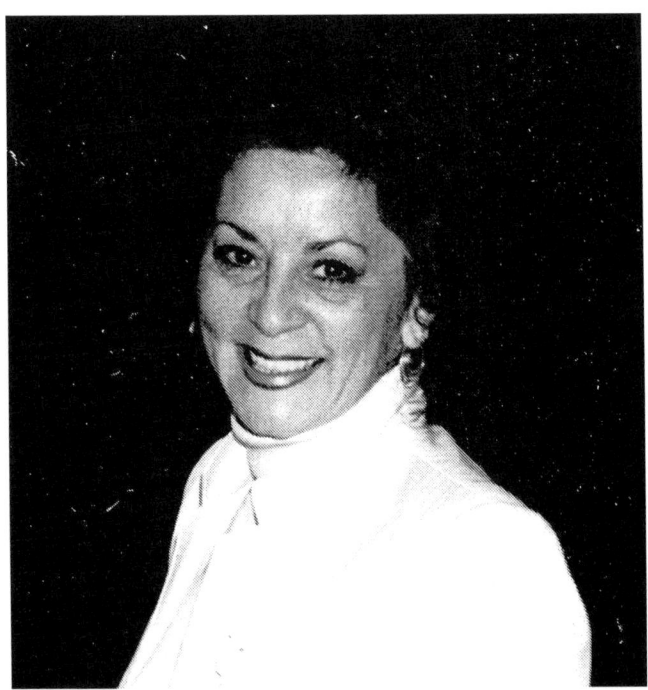

Rikki Barton, volunteer worker and air racer. Creative at crafts, she also teaches Spanish dance.

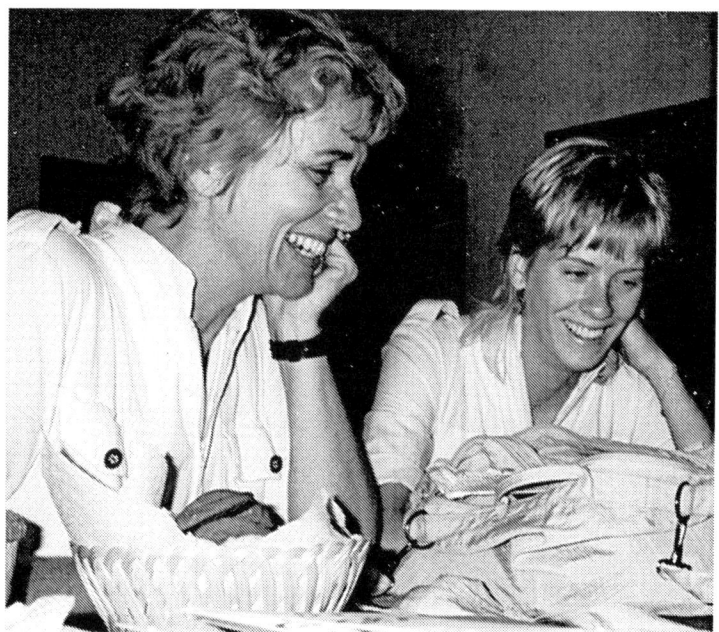

Sharon Crawford, major airline jet pilot, shown with her daughter, air race co-pilot and veterinarian, Carolyn Crawford.

Donna did a great job with publicity in Oregon, working with newspapers, television and radio from publicity releases sent by my sister Betty, race publicist from 1984 to 1994.

The city of Bend was especially excited when its own team entered in 1986. Mary Jo Grandy and Mary Ann Campbell, though first time racers, took seventh place. First place went to airline pilot Sharon Crawford and her daughter Carolyn.

Bend was the terminus from 1984 untill 1996. They treated us royally. Unfortunately, a woman bought a ranch next to the airport, added alpacas (similar to llamas) and when the 1996 race ended at Bend, she filed a complaint about the race planes coming too close to her ranch, upsetting her animals.

The FAA at Portland, Oregon, suggested Prineville, about 15 miles further northeast from Bend, as an alternative. Donna Vasko suggested we talk with airport manager Buck Taylor and town officials. We did, and both the 1997 and 1998 races ended there, and so will others in the immediate future.

Prineville is a small town with inventive people working out their own solutions to problems. For instance, when the railroad came through Bend, Prineville learned it was not to be on the line.

The town built its own line to Redmond to carry produce and passengers. Today they also use it for scenic rides for groups such as ours.

A women's organization, the Soroptimist Club, owns the Senior Center and rents it to groups. It caters our barbecue dinner and our Awards Breakfast. Their husbands drive us from the airport to the hotel in vans owned by the Soroptimists.

We pay for these services and for the Center. We love it when we find enterprising groups like those in Prineville.

Crater Lake Flyers Chapter members from Klamath Falls do the timing at Prineville, led by Rhea and Richard Bastian.

Santa Monica Airport has always been the Start point of the race. Distance to the Finish is approximately 750 miles. We stop overnight at about the two-thirds point. This was Red Bluff, California, for many years, then it shifted to Redding, for several years; currently it is back at Red Bluff. The Red Bluff RON was first hosted by the Sacramento Chapter and then by Mount Shasta Chapter when it was formed in 1975.

There is one more official timing point between Santa Monica and Red Bluff, and this, too, has changed through the years as airport conditions or officials' attitudes shift. The airport is chosen to meet the fuel needs of the smaller planes. First it was Chowchilla, followed by Merced and Modesto. Currently it is Castle Airport.

Some of the Powder Puff Derby rules adapted for my race included daylight flight only and under visual flight rules. Safety is always the number one consideration.

Aircraft are family-type, two- or four-place Cessnas, Pipers, Grummans, Beechcrafts, Citibrias, Aeroncas, and single or twin engine. Each is handicapped, with the winner averaging the highest ground speed in relation to its par speed.

Each pilot competes against her aircraft model's par speed and not against the entire field. What really counts is how skillfully she uses her knowledge of flight weather, navigation, radio work and the flight characteristics of her aircraft.

Most racers come from California. Once we had a mother-daughter team from Texas. Other states represented from time to time are Arizona, Nevada, Utah, Washington state, and Oregon.

Invaluable volunteer help comes from several Ninety-Nines Chapters: Palms, Los Angeles, San Fernando Valley, San Joaquin Valley, Mount Shasta, Central Oregon, and Crater Lake Flyers.

Each year, the San Joaquin Chapter prepares a most delicious buffet lunch to those dropping out of the sky at the mid-timing

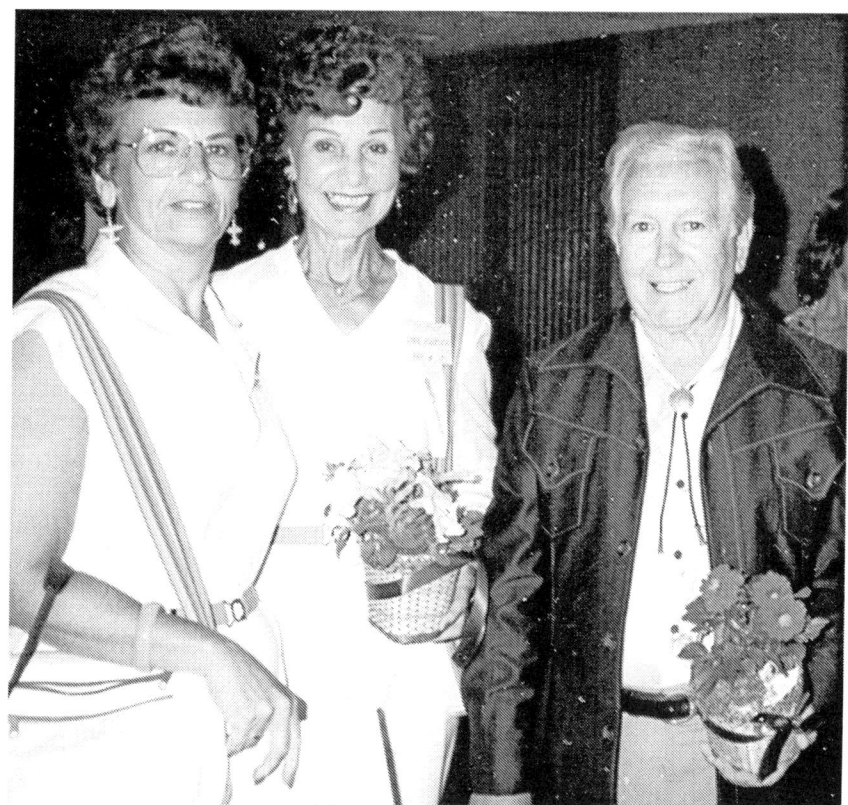

Race team Mary Bovee-Chesnut and Lorene Robertson with H. Glenn Buffington, writer and active supporter of women in aviation.

point. Some racers land just to enjoy the buffet and to say hello to friends.

Mount Shasta members handle timing, hotels, transportation, publicity, and whatever else is needed at the over night point.

Most of the Mount Shasta members live in the Redding area and several highly competitive teams from there race each year. Teams such as Donna Taylor and Jane LaMar will be in or near the winner's circle, with Suann Prigmore and Lois Van Zelf hovering close or beating them to it. These teams are usually in the top five and, while friends, are fierce competitors in the air.

Barbara Boot, who races nearly every year, does a great job with publicity in the Redding-Red Bluff area, as do those named above. All are quick to use any publicity opportunities.

Shirley Leatherwood handles the City Park festivities and the timing, transportation and publicity in Red Bluff.

Publicity is not about getting famous or a sponsor, it's about keeping up interest in aviation: keeping that airport open, getting children excited about airplanes. Most women pilots know this and try to work with the news media.

In 1985, Eleanore Scott of Trinity Center, near Redding, co-pilot for Beverly Romero, wrote me that she was amazed how many people told her they had read about her and the race. This is part of the fun, to have friends know about and vicariously enjoy our activities.

Entry fees pay all race costs, trophies and prize money. Prize money is limited to five hundred dollars total divided among the first five places. Prize money is deliberately kept low so that the focus is on the fun of flying. First place money just about pays fuel costs for a single-engine plane.

Place trophies are awarded to the first ten places, with identical ones for pilot, co-pilot, and passenger. The identical trophies acknowledge that a PIC (pilot-in-command) may be a private pilot with minimum hours, while her co-pilot may have thousands logged as an instructor. Some instructors often fly as co-pilots for their former students who have their private certificates, so the new pilots can experience that special camaraderie enjoyed by women pilots and can also be the PIC. Leg trophies go to those with the best time between timing points.

We have special trophies for relatives' teams, best time for first-time racers, best time for the team with under 500 flight hours. The last place team usually receives a "scenic tour" prize.

Some racers, knowing they will end up in the middle of the pack anyway, deliberately go after the last place prize. I went after it once when I knew I was close to getting it, but somebody 'beat' me to it. Beat out of last place!

We've had many mother and daughter teams, some grandmother and granddaughter, sisters, aunt and niece and other such combinations. Betty and I flew twice as a twin team.

After friends Dolores Pynes Hamilton and Dorothy Walker had flown the race four times and won it three times, each took a daughter, Lori Brown and Cheryl Madden, along for the next one. As they expected, four in one plane slowed it so they didn't win, but they had such fun together they did it again the next year.

In 1993, corporate jet pilot Mahlone Becker and co-pilot Robin Becker, her mother, raced. Mahlone had taught her mother to fly and I, as the FAA flight examiner, had given her the private certificate.

In a later race, Mahlone and Robin were joined by Robin's mother, Vera Ives, then 89 years old, and Robin's sister, Fidge

Brown. Vera was ill the next year and deeply disappointed she couldn't race again; her place was taken by Mitzi Brown, another sister of Robin's.

An Arizona race pilot, after meeting Vera Ives, decided to take her 91-year-old mother-in-law the following year. She did so, and from all accounts they had a lot of fun.

In 1991, my co-pilot was granddaughter Michelle Brittingham Fox. Shortly after she was born I predicted that she'd be my co-pilot in 1991 at age 18: she was such an adventurous baby. We had a great flight together.

Jean Larson Schiffmann, my friend now for more than 50 years, flight instructor, mother of four handsome sons, has entered nearly all of the Palms to Pines, nearly always placing in the top ten, and most of the time, in the top five.

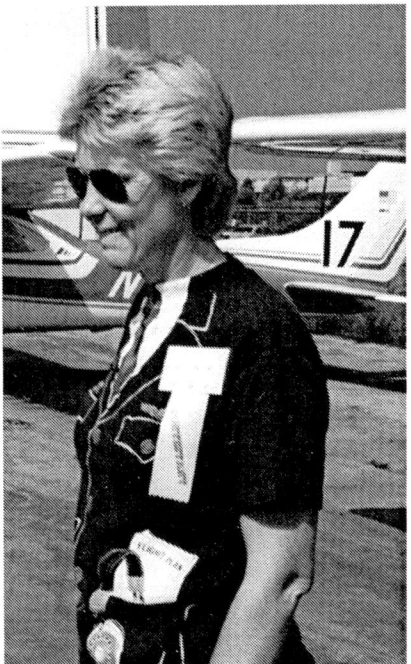

Diane Winn, active race pilot, is ready to go again.

Jean's late husband, Charles ("Pat"), was an FAA inspector instrumental with establishing FAA air show and air race policies. He worked with our aircraft inspection teams in the last years before his death. Jean says that sometimes other race pilots asked her if that wasn't a conflict of interest, having her husband a race official.

"Not at all", Jean told them. "He wants me to win as much as I want to win."

Jean's nickname is "Sunny", and it describes her; she is fun to be around. It was she who kept saying, "It's the most fun race of all!" Betty added that to all of her press releases, and everyone repeats it in all the press interviews, and makes it so.

Los Angeles Chapter members Norma Futterman and Aileen Pickering were a team in Norma's Beech Bonanza for many years. They always wore identical and distinctive clothes. They were always ready to help the race. In 1986, they generously took a Los Angeles Times reporter as a passenger.

Norma and Aileen kept the reporter informed and gave her the thrill of a good buzz over the timing line with other race planes coming in. There is nothing like it.

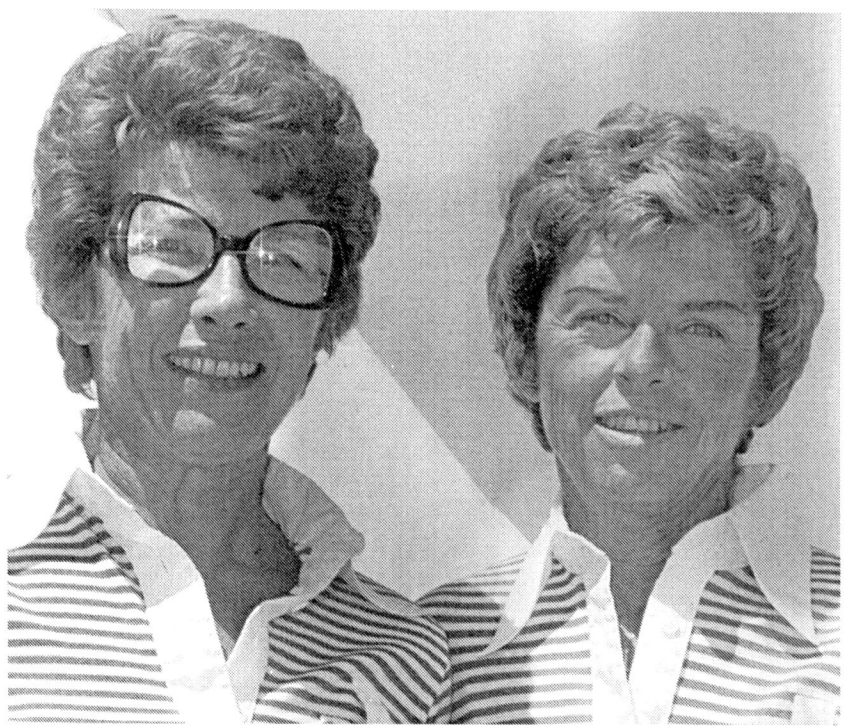

Norma Futterman, pilot (left), and Aileen Pickering, co-pilot (right), have flown in many races, and have been of considerable help in the Palms to Pines.

Another Los Angeles Times reporter quoted Esther Grupenhagen about her racing: "Basically, I take off and fly like hell and then, if I can find the airport, I look for something that looks like a white cupcake with an upside down ice cream cone on top [a navigational device]. Then I say, "Hang on, Helen, [Hewitt, her co-pilot] and we both go 'eeeeeahh!'"

Esther added that she likes to improve on the thrill of the 200-foot above the ground fly-by at 230 mph in her Bonanza by finishing with a double roll she describes as a "whoopdedoo and a whickadilly." She does this, of course, well beyond the airport over the "boondocks".

Recently, Norma and her friend, Maribel Llorens, climbed the many stairs to the Santa Monica Tower and worked with my sister Betty as Start Timers. Then they descended the stairs, got into Norma's Baron, while I climbed in back as a passenger, and we flew off to beat most of the race planes into Red Bluff, our RON.

When Lillian Camden was in her middle 40s, her husband Conrad says he read a story about me in a magazine and

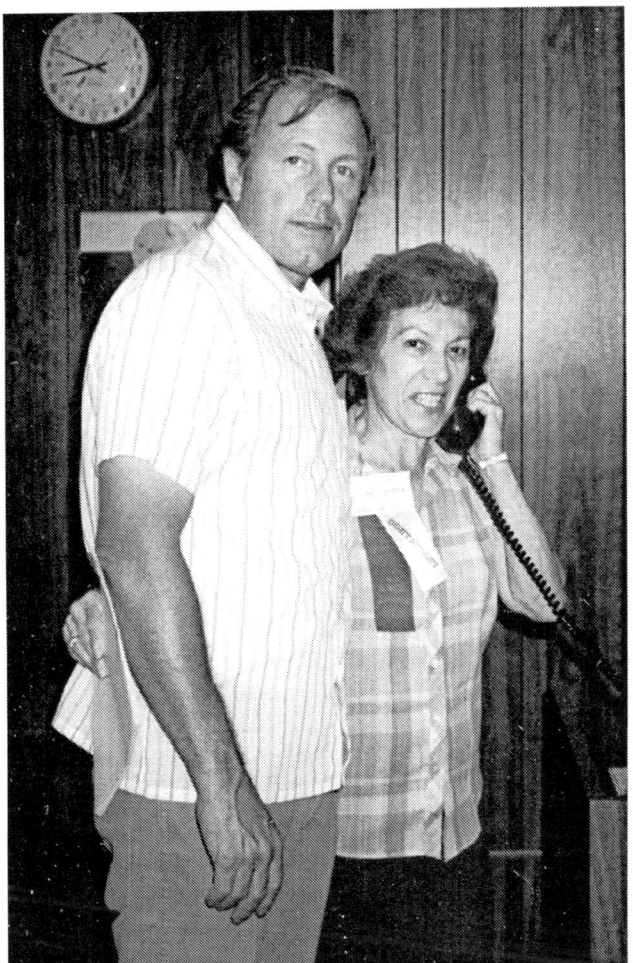

Conrad and Lillian Camden helped with the Palms to Pines for 20 years. She in various capacities, he with a mike in hand.

remembered that his wife always liked airplanes. He bought her a one-hour flying lesson and, as they say, "that's all she wrote." She learned to fly, and when she got her license, so did he. They bought a Grumman Tiger, she joined my Palms Chapter, and was one of my race committee members for the next 20 years.

Conrad, always supportive of his wife, handled the radio ommunication chores at the Start and in many other ways made himself invaluable to us.

When Lillian and Conrad retired from racing in 1994, Jeff Weldon, husband of Palms member Cecilia Weldon, took over the

Start radio communications work. Cecilia became our race/Chapter treasurer and my right-hand assistant in all things.

The other member of my race committee was Nina Laughbaum. Like Lillian, she learned to fly after her children were grown, bought a Piper Cherokee, and joined my Palms Chapter. She and Lillian were a team for the next 20 races, switching to the other's airplane every other year. Nina still races and assists her daughter Marilyn Laughbaum at the Finish in computing times.

Nina's friend, the late Harry Jugend, helped with many race chores. Whatever Nina wanted, Harry did it. He helped inspect race planes and helped timers in the Tower. Then he and Conrad would climb in whichever plane was left behind and fly to the terminus. They were welcome; they did not depend on their gals to look after them, nor did they critique race performances as some husbands did at first until I stopped that destructive practice — at least, in my hearing.

For awhile, husbands were not welcome unless they were part of the race committee and had a job to do. Most of them are terrific and very helpful.

My committee handicaps planes, checks paperwork of entrants, inspects planes with other helpers, collects fees, pays bills.

I work with the FAA, the airport tower personnel, airport managers, hotel managers, hotel dinner caterers, and transportation. I oversee all race matters, make decisions and resolve conflicting situations.

Whenever I need help, someone will step forward. Two special chores are on the revolving list: tower timers and the honorary starters who wave the green flag as each plane takes off.

The timers at Santa Monica are usually dedicated to the job. Three are needed in the tower. One to watch each plane pass a certain point on the runway, and two to get that time down to the second. They check each other.

Sometimes a pilot who can't race that year helps. This was the case with Doris Minter, Emma McGuire, Robin Becker and Ann Armstrong in various years.

Mary Francis "Fran" Blair helped time, faxed the results to the terminus and then, some years, hopped in her motorhome to drive to the Finish just to be with the gals. Once she arrived at Bend, Oregon, to find that the fax with the race times was discarded when it was received by a hotel clerk. Fran had another copy with her. Once a flight instructor, later on a high school teacher, she was well organized and always eager to help.

We've had problems: once a timer came early on the morning of the race start, decided to do some jogging, but forgot to come back. Another time the same gal forgot to come at all; she called two weeks later to inquire if the race was coming up soon.

One year, two of the three timers, laughing and giggling, refused to go to the tower until an hour after the race was to begin, despite the hot weather and sweltering people in planes, ready to go.

Over the years, in all but these few times the timers have performed as they should, but, unfortunately, the weather often does not: huge fog banks and poor visibility in all directions delay the start.

Once we had a tower operator delay the race, making up his own rules; making each pilot request permission to take off, holding each race plane on the ground while touch-and-go-around-the-airport fliers did their thing. The operator hurt all the traffic badly that day, and he was not allowed ever again to have anything to do with the race. Since then, Pam Choi has taken personal charge of the race day, runs a tight tower, and all goes fairly well at Santa Monica.

We have a professional weather person give the forecast the night before at the pilot's briefing meeting, and again at the airport on race morning. Currently this is Stuart Macofsky. He has also served as a timer twice.

At refueling and the RON airports, the timers sit at a table placed near the runway the pilots are to buzz. Usually special watches are used.

To help combat any problem, each pilot is also asked to keep her own time in the cockpit, so that her records can be consulted when a puzzling entry appears on the official time record sheet.

Each year we have a celebrity wave the green flag. We've had movie and TV actors Robert Hays, Susan Oliver, Justine Bateman, Estelle Getty, Emily McLaughlin, Prince Hughes, and Los Angeles area television news anchors Jann Carl, Marta Waller, and Dr. George Fischbeck. I had given Private certificates to both Robert Hays and Justine Bateman.

Our own Palms Chapter member, Ann Savage D'Armand, pilot and charming movie actress who was "Queen of the B Pictures" in the 1940s and 1950s, waved the flag for four race starts at Santa Monica, 1993 through 1996. She, herself, has raced twice.

Romance novelist Judith Krantz was both our 1987 dinner speaker and our flag waver. Gracious and funny, the quick-witted and beautiful author spoke to a roomful of adoring fans. She posed for photos at the dinner and again the next morning at the airport with her husband, Steve Krantz.

Judith had spent many hours at my home the previous year interviewing me on aviation matters for her book, *Till We Meet Again*. I was just one on a long list of aviation experts to be consulted by Judith, but she was quite warm to me on her Acknowledgement page in the book. What fun!

Usually all goes well with our Palms to Pines Air Race, but sometimes it doesn't. We adjust or change rules, and race again the following year. The idea is always to enjoy the sport of flying as well as the special camaraderie among women pilots.

Diann Laing wrote the following story about the 1990 race. She gives us permission to print it. It was in the September/October 1990 issue of *The Southwesterly* newsletter of the Southwest Section of The 99s.

Diann: "This year's Palms to Pines could have been called the Fog to Smoke! Count on Claire Walters to try a different look to spice up tradition!

"Lisa [Schilling] and I flew down from Reno, Nevada, the Thursday before the race. The western side of the Sierra Nevada was marked for miles with pillars of smoke that fanned out and settled into the mountain valleys. We stayed about 12,500 feet to keep reasonably blue skies. However, when the Los Angeles basin emerged on the horizon, it sadly looked like a gray smudge. Fortunately, it wasn't smoke but I wasn't sure that fog or smog would be a great improvement. We picked up a clearance over Porterville and headed in.

"The bustle at the check-in table was as friendly as ever. Good to see familiar faces and feel the excitement in the air. We joined fellow racers in the ride to the hotel and settled in to the racing chatter and planning. Lisa and I were determined to benefit from our previous race the year before and started hounding anyone who could give us the insight on picking a route and altitude: through the pass? over the mountains? low? high? climb fast? climb slow?

"Friday dawned grayly. There was no horizon; nearby buildings were blurred. So we all hung out and looked daggers at the scud sliding slowly inward from the coast. We all willed it away — a peek of sun to the east! Promise of more...maybe. Finally, about noon...pilots to planes! We'll depart to the east on runway 3, left turn over the freeway and head north.

"Strategy time: Shimoda MO (my Cessna 172) was trying her first race. She is not a climber — just can't train her. So Lisa and I decided to stay in the canyons, stay in the trees. We came screaming out of Gorman. I looked apprehensively at the redline. We went down on the deck in the haze and heat. If low was where it was at, we would be low. It was kinda neat to look at the tractors from what appeared to be eye level.

"Lisa was chasing CDI dots with determination. At one point the LORAN lost its signal. Not forced to go to pilotage! Oh, no!! Master signal captured again. Whew. We were pleased with everything and feeling good near Modesto when a Warrior went gliding by as if we were standing still. I saw race numbers. Depressing. We made a good fly-by at Modesto and stuffed ourselves with the great fruit, cheese, and nutbreads the San Joaquin Valley 99s provide there. A welcome treat in the cool and shade.

"Back on the fly-by take off. What to do about the Sacramento ARSAs. Go over? Go through and risk working with them and getting vectors? We decided to listen to the controller and see if he sounded like a decent type. He sounded marginal. We were at a 1000-foot AGL [above ground level]. Our course was through a narrow channel that wouldn't compromise McClellan or Metro [airports]. Called him up and tried to sound clear, professional, and polite. We explained we didn't want vectors. He sounded slightly miffed and he gave us a squawk.

"We were still at 1000 feet AGL. Level. Approach called and told us to climb to and maintain 2600 feet. Lisa and I exchanged looks. We started a climb: very, very slowly. Maybe ten feet an hour. He called us back. How high? We told him we were doing a slow climb because the engine was really hot. He asked if we were having trouble. We said no, just keeping things cool.

"We were handed off. Just a bit further. Still at 1200 feet. We fudge some more about climbing, then we're out! (Our efforts earn us a plus 18 for that leg.) On to Redding. The foothills to the east were remotely distinct in smoke. Terrible fires beyond Chico. Thank goodness the wind was from the southeast. Redding was wonderful: pool and barbecue. Great to relax and swap stories after such a challenging day.

"The flight the next morning to Bend was sobering: smoke had settled in the valleys and canyons. Some of us picked our way through on the deck, some went high. We were within five miles of Shasta and never saw that towering mountain. Bend is always unsettling to find. Lava domes everywhere and nothing remotely resembling an airport. Moment of doubt. Suddenly: Tally Ho! and we're boring down on the field. And MO screamed by as best a 172 can. And it's on to a river trip, or swimming or fishing or... just good conversation.

"We finished in the top 90 percent. Lisa, Mo, and I had a great time. We'll be back next year...and maybe try the high route. At least we'll see Mount Shasta!"

Jean Schiffmann has been in nearly every Palms to Pines, as well as several PPDs and Air Race Classics. She entered our first race in 1970 in a Cessna 150, a nice, stable aircraft, the capabilities of which she was quite familiar. She'd flown solo, with no problem, a "finger on the chart" pilot.

The following year, she drew number 13 for her aircraft and added a brand-new private pilot to fly as her co-pilot, Karen Sherman. They were a team in many more Palms to Pines races and in the 1975 Powder Puff Derby.

A raconteur of considerable talent, Jean tells the following story to audiences, which she describes as the most humiliating experience of her life, certainly of her 21 years of flying; but with time, the funniest.

Jean: "Race plane number 12, a twin-engine Piper Aztec, piloted by Jan Gammell and Ilovene Potter was next to us, and would take off just ahead of us. I had borrowed a Citibria, a low-powered single-engine. I said to my husband Pat that number 13 might either get blown over or airborne when number 12 took off. Pat turned to Jan and Ilovene and asked with a straight face, 'Couldn't you attach a cable to number 13 and tow them along? They wouldn't be much drag.'"

"Neither Jan nor Ilovene were amused — they didn't even smile.

"My airplane was a Citabria, the lowest power ECA model, a tandem airplane in which one pilot sits behind the other, and with a stick control and a tailwheel. To match its orange and white paint scheme, Karen and I dressed alike in orange shorts (they were called 'hot pants' that year), white shirts and white cowboy hats. Along with our big number 13 we painted a black cat on the tail. For luck, right?

"There was no Pilots Operating Handbook for this borrowed airplane, so how well it would climb, how fast it would go, how much fuel it would burn and how much fuel it held were all unknowns. We quickly learned that climbing was not the favorite activity of this airplane so we had a very good view of the low-land scenery and a perfectly fabulous ride the first day of the race.

"Knowing that I would need both hands to fly this airplane, I had made a strip chart, four inches wide, of our entire route so I could hold the chart and control stick with one hand and the throttle and trim with the other. Anyone should be able to stay on course within the terrain shown on four inches [32 statute miles] of a Sectional chart, right?

"Day two: departing from Red Bluff on a gorgeous Saturday morning, I looked ahead to the Trinity Alps which we would have to climb over to stay in a straight line on the chart.

As the mountains loomed larger and higher, my mind went back to the day I had asked the airplane's owner how the Citabria performed at 8000 feet and his horrified 'It's never been to 8000 in its life!' remark. At the time I had dismissed that as nonsense, but it was becoming very apparent that we were not outclimbing the terrain, even with just one stopwatch, one full chart, one credit card, and ourselves in the white shirts, white cowboy hats and orange hot pants. Nothing to jettison.

"The airplane had only the basic required instruments, no navigation radio, and as we had no headsets, communication between Karen and me was difficult. I had learned that the airplane would not stay trimmed to fly fast without one hand on the trim continuously, and of course, the other hand on the stick.

"Making a sudden (and necessary) decision to alter our course and follow the freeway up over the pass to Ashland, I leaned back to try to shout at Karen to hand me her chart; we had now flown off mine.

"Whoops, why do I see only sky? Push the nose down!' Unable to hear me, my co-pilot leaned way forward with a 'What?' which, of course, filled the windscreen with a close-up picture of the highway. Now we became like a bucking bronco — nose up — nose down — nose up — as I tried to grab her chart.

"Unable to let go of the controls with either hand, I shouted 'Can you keep the wings level while I look at your chart?', to which Karen shouted back, 'I can't keep the wings level because I can't see the instruments! Your head is in the way.'

"The trucks were getting very large as I screeched, 'There aren't any instruments! Just look out the sides and don't hit a tree or a truck.'

"Realizing that I was not going to get any help, I gave up on the chart and just followed the freeway. Having passed the highest mountains, my memory told me that we should now be in a valley and approaching Roseburg, our next stop. But right in front of us was another mountain. Quick — which way to go around it? Shouting some more, 'There's a wide cut through the trees. That means there should be a power line shown on your chart! There's a bend in the road that we're following with a bridge across it. Can you find that?' Receiving nothing but negative answers to both questions, I realized that perhaps chart navigation was no longer stressed as much as it was when I first learned it.

"By now I could smell the ocean, and since we were not supposed to be anywhere near the ocean, I shouted that we'd have to turn around and try to find an airport as we must have missed Roseburg. Since there was no longer any hurry — why hurry when you don't know where you are going — I trimmed the nose up to go slower and the engine sputtered. Thinking that Karen must have bumped the throttle to idle, I asked her to please not touch the throttle, to which she replied that she hadn't.

"After trying again to trim the nose up, with the same result, I reluctantly announced that it appeared that we could only fly downhill from here. At this point Karen said, 'There's a green field down there. Why don't we land there?'

"'I answered, 'Why would you want to land there? There's no house or any buildings.' When she replied that there was a stream there and her feet were hot, I determined silently that this plane would fly if I had to hold it in the air with my teeth.

"Proceeding inland, in a gradual descent, we presently saw more than one airplane in a northbound direction, which meant they were race airplanes. More importantly to us, there was a green field with an airplane in one corner. Must be an airport! It was bounded by a sawmill, a lake, a town, and the inevitable power line.

"Reverting to my emergency landing training, and knowing that we could only fly downhill, we made an uneventful landing. Yes, there was a fuel pump here, but locked and no directions for its use. Nothing to do but hike to town. Easily said, but first there were blackberry vines, a barbed wire fence and a creek to cross.

"Not being choosy, we knocked on the door of the first house we came to. On this beautiful, sunny, warm Saturday, around nine or ten in the morning, the next moment is etched in my memory forever. The woman who came to the door could not have looked more astonished if we had been from Mars. Never had I been lost or even had to land anywhere I was not intending to go. I was so chagrined that I could have dug a hole and climbed into it.

"Instead, nearly gagging on the words, I politely inquired, 'Can you please tell us where we are?'

"'Her reply was, '553 D Street.'

"'And what town would that be?' She named a town, totally unfamiliar, so I had to ask if that was anywhere near Roseburg. She replied that Roseburg was about ten miles down the road. As I was about to ask if there was some way to get fuel at the airport, her phone rang. There was nothing for us to do but wait politely. A short lifetime later, 45 minutes actually, she finished talking and was able to tell us that the airport manager lived down the street in the green house.

"When we knocked on the door of the airport manager's house, 14 little heads appeared in the large picture window. A very sleepy-looking woman appeared next (seems there was a sleepover party) and after another short lifetime, an even sleepier-looking man appeared, dropping into a chair with coffee in his hand. Being a coffeeholic myself, I understood there was no way to rush him, and besides, by this time I had given up on completing the race.

"As he drove us to the airport, I asked for ten gallons of fuel, enough to get us back to Roseburg for our scheduled stop. But when he stated that there was no gauge on the fuel pump, it occurred to me that we could fill our fuel tank, and overfly Roseburg and possibly still beat the deadline to the Finish at Independence.

"Since we had actually flown about 40 miles past Roseburg, we had to fly back to the designated reporting point, calling in for a - 'Fly-by to continue'. After a long pause, Roseburg answered, 'Ahhhh, Race 13, verify that you are NOT stopping for fuel.' On we flew, arriving at Independence 15 minutes before the deadline. We were besieged by the newspeople wanting to know where we had been and why there was dried grass stuck in the tailwheel. 'Would you believe we stopped to see an old cropduster friend?'

"The sad part is that anything that can be construed to be a problem, or better yet from a newspaper viewpoint, a tragedy, gets the most publicity. Race 13 was on the front page of an Independence area paper. The funny part is that some unknown person got more lost than we did and won the 'Scenic Award' for the one who saw the most scenery. We were number 42 out of 43. After which my co-pilot announced, 'Jean, you not only can't win, you can't even lose right!' However, we did win the prize for the first leg.

"The happy part is that after this experience, Karen insisted to me that I should be an instructor, saying, 'You have taught me more on this race than all my other instructors put together.' And so I did. I became an instructor, and she became the best navigator you'll find. And that's what racing is all about: learning, making friends, meeting new people, and having lots of fun and memories."

The following ditty was composed by Norma Futterman and Aileen Pickering and sung at the 1988 race. Norma strummed a guitar as everyone sang to the tune of "On Top of Ol' Smoky":

The Racer's Lament

Come all you brave racers and listen to me.
I'll tell you a story that starts by the sea.

Took off Friday morning on runway two-one,
Right turn at the shoreline, were we having fun!

The palms were a waving. The weather was great.
Santa Monica traffic would just have to wait.

Claire told us, 'Don't flat-hat, the neighbors will shout!'
What she means by 'flat-hat' we can't figure out.

We crossed over Gorman, our airspeed seemed high,
But a Cessna 150 just passed us right by.

We flew by Modesto, avoiding the heat.
We're heading for Redding - we'll have time to eat.

Off Saturday morning. We followed our lines
Across all those mountains all covered with pines.

Alongside Mount Shasta, all covered with snow,
We lost our mag compass - Know not where to go.

Called in abeam Dorris - old habits die hard.
The tower at Klamath was caught off its guard.

R'member Independence - for years 'twas the end,
The fog drove us inland - We now land at Bend.

They hand out the prizes - Lil, Nina, and Claire.
Rikki keeps them honest - The race must be fair.

We pray as we listen: 'Please call out our name.'
The 'Scenic Route' trophy is our claim to fame.

*Palms to Pines Air Races has just raced its 30th year.
Hundreds of women pilots have enjoyed it over
the years, and we are sorry we have so little
space left and can only print
a few of their pictures.*

Chapter 19

It began quietly and calmly enough, like the butterfly in the jungle that flutters its wings, causing air to move, and a storm develops in the moving air somewhere far away.

In 1994, our Palms Chapter learned that a nephew of Mathilde Moisant, a 1911-12 aviatrix, had stored his aunt's flying suit and other artifacts in his garage and he was moving to Lexington, Kentucky, with them. His name was John Weyl.

We asked him to give the artifacts to us and we would ship them to The 99s Headquarters Museum for display in a case we'd have built in Oklahoma City. He was reluctant to do so but finally agreed; otherwise Mathilde's things would remain in boxes stored somewhere. The artifacts, besides the flying suit made of green corduroy, which was designed and sewn by Mathilde, was her pilot's license number two, goggles, binoculars, and boxes of photos, letters, and news clippings. All were donated by her nephew John Weyl and grand nephew, David Zeitlin, a Mar Vista (Los Angeles) resident.

Our chairman that year, C.J. Strawn, is a movie production designer. She designed and ordered a cabinet to be built in Oklahoma City to house the collection. We found that a child mannequin was the only size that the flying suit would fit, but the helmet and boots were too small. Rikki Barton shaved the head and the feet of the mannequin so it could be fitted with them. Rikki then packed the box for shipment to Oklahoma City. Problem: the box was too big for ordinary shipping, so we had to find a shipping company for the job. We did, and it was very expensive. The box arrived there two weeks before we went to Oklahoma City to present Mathilde's things to Headquarters.

Palms chairman C.J. Strawn, vice-chairman Gail Kass, and I flew via Southwest Airlines to Oklahoma City to attend The 99s Open House in March 1995. There was no Palms Chapter display case; it had not been built. The person hired to build the case finally showed up and said he'd have it done in 30 to 60 days. C.J. drew a new design for him. We planned to make a five-minute video tape and place it in a monitor by the case.

We asked Loretta Gragg, the then Executive Secretary for 30 years for The 99s, where the case would be placed.

She said, "On the second floor when the museum is built."

We asked when it would be built. She said that no one knew. When the building was constructed in 1988, a second floor was added for a future museum the officers now spoke of so vaguely. No timeline was established for it and no money had ever been set aside for it even though several estates had designated money for a museum.

We asked about the account on the books which was established for various items a member might want to contribute to. The museum was one item. Over the years, a total of $9,000 was accumulated in it, we were told.

We left our boxes on the second floor, alongside many other boxes of memorabilia, and flew home with heavy hearts. We reported to our Chapter what we had found.

We sold raffle tickets during the next Palms to Pines Air Race to help pay for the $3500 display case, video tape and monitor expenses. We had the money, but the case was not built. At our

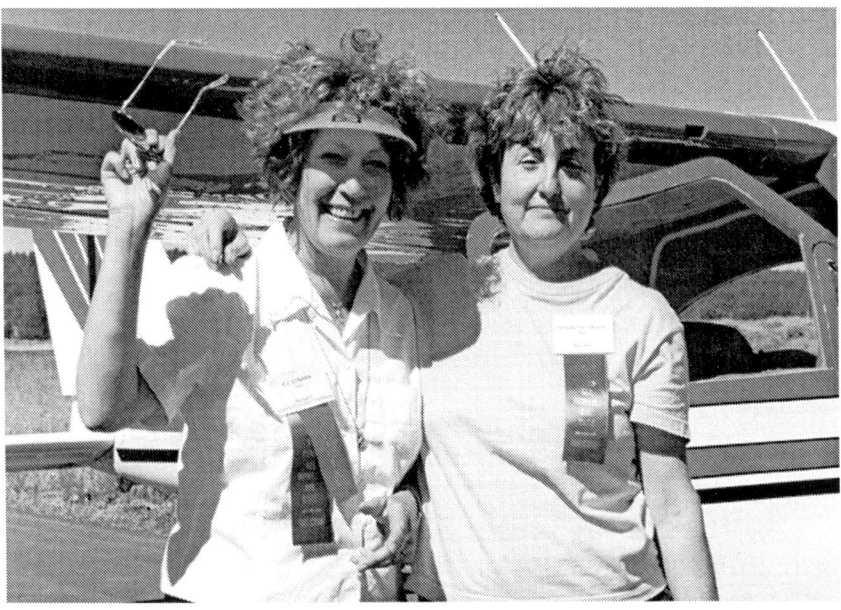

Designer and construction overseer of the museum was C.J. Strawn (left). She spent three years on the project. Susan Theurkauf (right) helped raise funds for it. Here the two are about to race in the latest Palms to Pines Air Race.

Photo courtesy of C.J. Strawn

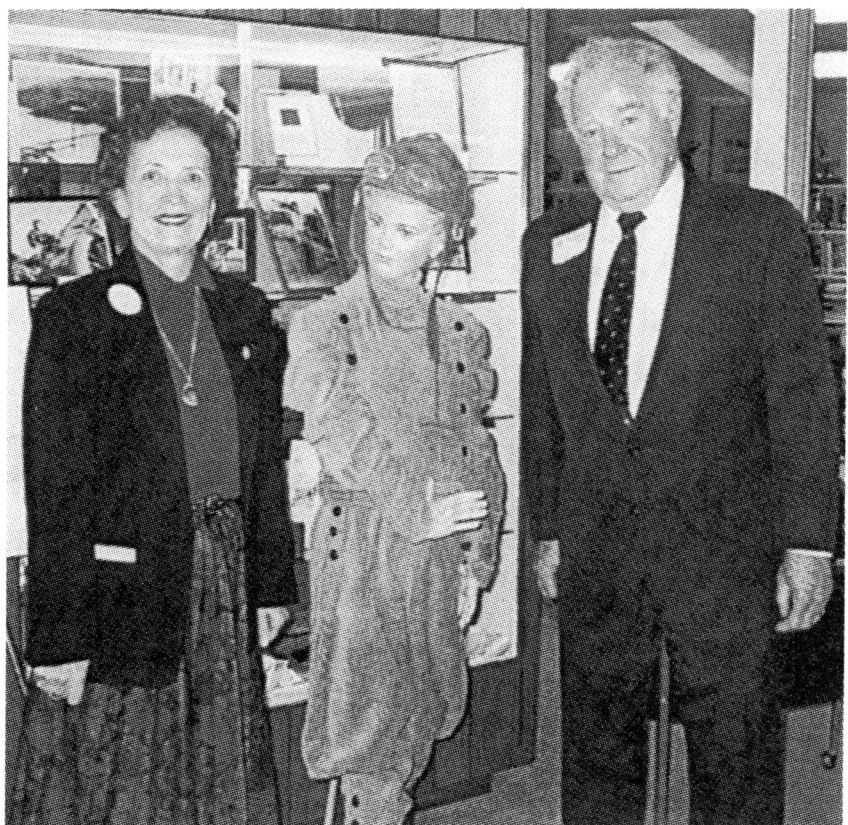

Joyce Wells (left) stands next to the mannequin dressed in the Mathilde Moisant flying costume. At right is John Weyl, nephew of Mathilde.

Photo by Peggy Ewert. Courtesy of The 99s Archives

Chapter meeting in January 1996, C.J. and I discussed the status of the "museum". It had to be built.

Neither C.J. nor I can remember who said it first, but one of us said, "Let's do it." She said she'd design and build the museum, and I said that I'd raise the funds. We donated the first of the funds, $500 each, knowing that we couldn't ask others for money if we didn't contribute. Our Palms Chapter treasury contributed $2000, and members gave another $4565.

I called all The 99s officers and directors at their homes, told them of our Chapter plans to build the museum, and asked them for their verbal okay. They all gave it.

C.J. researched extensively several museums in the Los Angeles area and then drew detailed plans for the interior design of the museum. We live in the Los Angeles area and we were about to

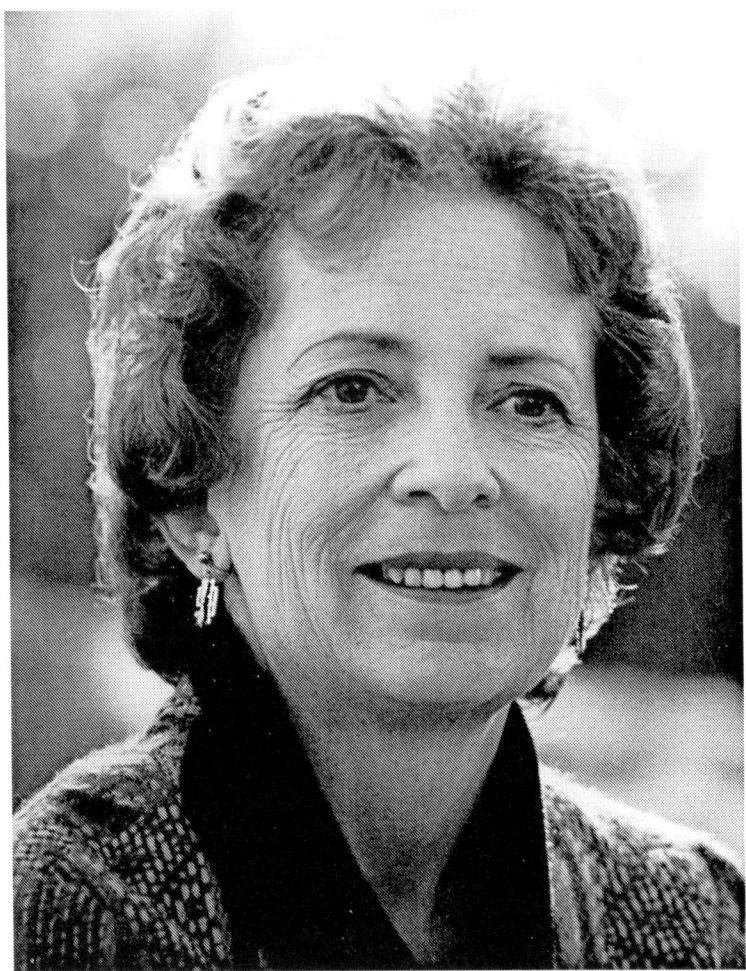

Joyce Wells, 99s President, 1994-1996, said the magic word: "Yes!" when Claire Walters and C.J. Strawn presented the museum proposal.

undertake building a museum in Oklahoma City. C.J. was quite enthused about the project; it was the first time she was building something permanent — all her other projects were designed to be torn down.

I typed the story of the museum-to-be with a request for money, added C.J.'s design, and printed 500 copies. I mailed these to all Sections, Chapter chairman, and my 99s friends. We attended the Winter Board meeting in San Diego and picked up a few checks.

I had already phoned everyone I knew to tell them of our plans and that we'd have a Wall of Wings with names going on it of those

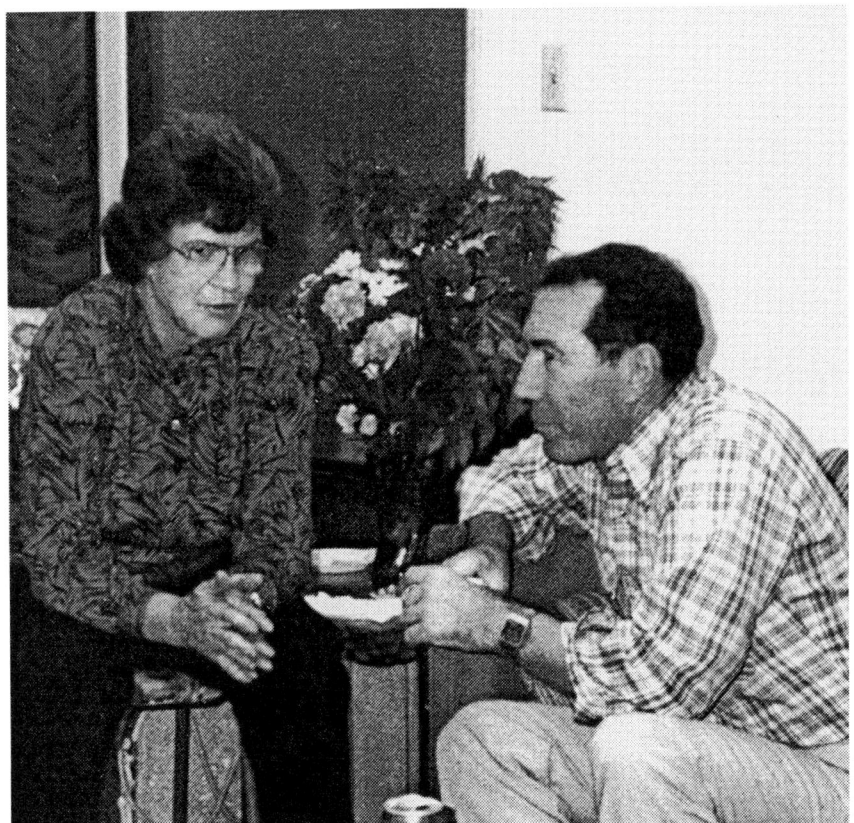

Shown here at Claire's "retirement" party in 1987, Mary Francis "Fran" Blair and Harry Jugend plan their next assignment as Start Timers. They always helped the Palms To Pines Air Race, doing their work quietly and well. Fran left a bequest to the museum.

Photo by Betty Loufek

who gave us $500 or more. And I was very sure all my friends would want their names there!

Also, for $1500, one could have one's name put on a cabinet. We learned later that a cabinet would cost between $1200 and $3500.

Originally, my naive thought was that if all Chapters donated $500 each and each member contributed $10, we'd have enough. One Chapter, North Dakota, with 11 members, promptly sent $110. What I didn't know was that, in the end, only five percent of the membership and 40 percent of the Chapters would donate.

The timeline C.J. wrote was in three phases. In Phase 1, we'd publicize the project and ask for donations to raise $60,000 by August 31, 1996. When that amount was raised, C.J. would fly to Oklahoma City and hire the construction people to start on the interior walls and electrical work. She'd select a furniture company who would build the cabinets out of wood.

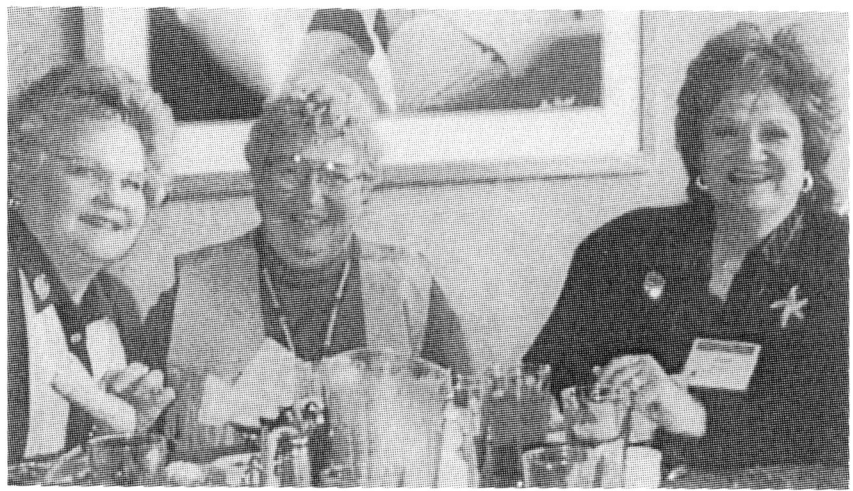

Left to Right: **Anita Lewis, Kay Roan and Jaye Howes. Anita and Jaye help raise funds for the museum, and serve on the Board of Trustees.**

Photo courtesy of The 99s Archives

Some of my friends sent nice checks; others asked, first, if I'd guarantee they'd get their money back if the project didn't start. They had sent small sums of money through the years to 99s Headquarters for the museum and, as far as they knew, the money had disappeared in some black hole. I am not wealthy, but I have credit cards; I personally guaranteed they'd get their money back if the project didn't fly. I knew we'd complete the project.

A valued member of our Palms Chapter, Fran Blair, had a terminal illness but came to our meetings whenever she could. She told us she was driving her motorhome to Washington, D.C., but would stop in Oklahoma City to see the projected museum on the second floor at Headquarters.

Fran died December 8, 1995. We learned then that when she stopped at The 99s Headquarters, she made a change in her will giving the museum project a sum of money. How much would be determined in probate, but we were told it would be about $15,000. A large sum to us, and quite unexpected.

It was March, 1996. At the Executive Board meeting in Oklahoma City, after I spoke about the museum and C.J. presented her plans, one director reminded the others of all they planned to accomplish and demanded that President Joyce Wells make a decision to either go with another project or ours. Joyce promptly said it would be our museum; the other project was not ready to start. To the director's credit, she was the first of the Board to send me a check. Joyce closely followed it with hers.

My sister Betty sent out news releases to all the aviation clubs, aviation newspapers and magazines, and to some aviation museums. She also sent letters over my name to aviation businesses and organizations asking for contributions. Two responded immediately, Jeppesen and AOPA (Aircraft Owners and Pilots Association). Thank you, again, so very much.

Most of the aviation organizations did not respond; those who did claimed that they didn't have any money. Since when was a yard sale or a wine-tasting event, or some other such fun thing...well, don't get me started. Unfortunately, I got the same response from most of our own Chapters. I was learning the sad lesson that all fundraisers learn.

But my friends had responded and the money was coming in. Two activists of the San Fernando Valley Chapter, Susan Theurkauf and Jaye Howes, persuaded that Chapter to donate $5000, and they appeared at one of our meetings to tell us. They also had a plan to shake money loose from other Chapters and individuals. The SFV

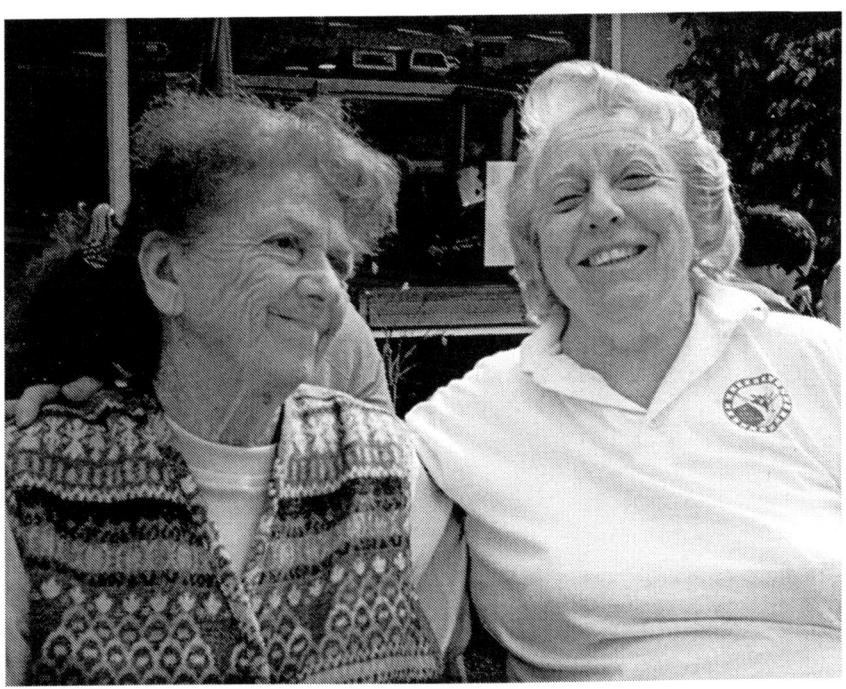

Emma McGuire (left) and Claire Walters talk over their nearly 40 years of friendship and the projects they have worked on. Emma was Claire's top flight instructor, among so many fine ones.

Photo by Shannan McGuire

Chapter is expert at raising money and at spending it on aviation projects, scholarships, all kinds of good things.

We put into operation the first part of their plan by attending the 1996 Spring Section meeting of The 99s to pitch the museum and ask for donations. Margaret "Peggy" Ewert, governor of the section, told me prior to the meeting that I could take all the time I needed. This was major — most speakers are severely limited. Peg was herself a contributor and was actively persuading others.

I gave my pitch to the Section, and on cue, Golda Neuman, chairman of SFV Chapter, stood up, waved a check for $2500 and challenged other Chapters to "meet it or beat it!" It caught them all by surprise, but many raced up to me with smaller checks and said they would do better next time. Competitive spirit is always high among The 99s.

This scene was repeated at the International Convention in Oklahoma City in July, where the other $2500 was presented by SFV to me, and the challenge reissued.

In the meantime, Susan, Jaye and C.J. had reserved the November/December 1996 cover of our 99 News magazine for the museum project. They got permission from Douglas Ettridge, artist, and from the painting's owner, Betty Bazar Robin, to use one of Ettridge's paintings on the cover.

Mr. Ettridge also agreed to create a painting to commemorate the opening of our museum and from this original work will be printed a limited addition of lithographs to be sold to 99s. This is still an on-going project as this book is being published.

Our fund's first stage goal was close to being reached, and with the growing momentum, C.J. and I knew we were going to make the museum a reality.

C.J. expanded and refined her detailed plans and prepared to fly to Oklahoma City to physically oversee the work start. I reached the $60,000 bench mark as scheduled, and C.J. flew away.

Betty wrote the first of several reports to our contributors announcing that fact. We have tried to keep our contributors informed at every stage — while asking them for more money, of course. Our Palms Chapter treasury paid for all this; we never used contributor's money for fundraising, phone, and mailings; all of the money received was sent directly to 99s Headquarters for our special museum account.

The Palms to Pines Air Race in August, 1996, gave us a chance to raise a little more. Robin Becker got prizes from aviation businesses around Santa Monica Airport, sold tickets for a drawing, and before she was through, had made over a thousand dollars for the museum. She and her committee have done the same since then at each race time. Robin is one of those who do what they say they will do, and on schedule.

One of C.J.'s concerns was that The 99s building is modern, built in 1988. How to give the museum a bit of early twentieth century craftsman look? She decided she might achieve it by going with linear oak and keeping the trim and molding clean and simple. C.J. spent time on each such detail, and for three years, now, has met each problem, solved it, and moved forward.

She flew to Oklahoma City frequently, and was almost in daily contact with one contractor or another by telephone and fax. This while doing her movie work, and flying privately besides. An iron woman!

My fundraising efforts continued, with many others adding their voices and checks: Anita and Vic Lewis, Margaret Callaway, Fran Bera, Vanecia Adderson, Betty H. Gillies, Wally Funk, Stacy Hamm Howard, Gene Nora Jessen, Jean F. Schulz, Donna Taylor, Dr. Jacqueline Boyd: the swell grew to hundreds of individuals, and all checks were welcome, from five dollars to many thousands.

As deaths occurred among our membership, memorials were created. While other memorial funds cost a great deal of money, our museum memorials were created for any amount, and thousands of dollars came in for the museum. Foundations such as the Hennings Memorial Fund made a substantial contribution. The brother of the late Ruth Nichols, a popular flyer of the 1930s, learned of the museum project and gave several thousand in her memory.

Memorial funds include Cynthia Karolich, Peggy Ewert, Betty Faux, Virginia Funk, Pam Van Der Linden. Page Shamburger and Vera Ives will also have memorials. There will be many others in the future.

As the money came in, C.J. proceeded with her work: had the contractors lay the terrazzo floor and the carpet, raise the ceiling in the library, paint the walls. Whatever there was to be done, she designed it, prodded the contractors, solved the problems.

In the meantime, we kept getting rumors about the settling of Fran Blair's estate. Apparently all had been done except for the bequests to a university and to The 99s. We had heard nothing directly and a year had passed. We didn't even know the lawyer's name who was handling the probate. It might take another year to settle. Finally the light dawned: we had a probate lawyer in the Palms Chapter — Gail Kass. Perhaps she could help.

I called Gail; she said she'd try to find out. Gail contacted the lawyer, told him about the museum project, and soon he concluded the estate matter as far as we were concerned, and the check arrived: $48,500. Wow. It took our breath away. Thank you, Fran.

The library will be named in memory of Mary Francis Blair and Dorothy Niekamp. Dorothy was The 99s first librarian in 1988 after the new building had been completed. She spent years doing research in the library, and important compilations of resource materials resulted.

In 1997 I attended two Section meetings, the Northwest Section, held that year in Victoria, B.C. (Canada), and our Southwest Section in Santa Maria. Both worthwhile. I did not go to the International Convention in Maine, but we had a friend, Jaye Howes, bring back our "Inspiration Awards", which C.J and I then received at a party at Jaye's.

In late 1998, we had enough money, it seemed, to finish the museum, and I turned my efforts to develop the fund to maintain and operate it. The money for that is coming in nicely, and it looks like we have successfully completed our project. As of January, 1999, three years after starting the fundraising, we have over $230,000. I'm going for $250,000 to insure the future of this wonderful museum.

I must pay homage to my beloved Palms Chapter. When I asked for 100 percent financial participation of the members, I got it, with donations running from $10 to $2116, for a total of $10,872. They have backed C.J. and me in all our endeavors. The Chapter treasury gave us $6250 plus paying another $1000 or more for postage, printing, so forth, plus Fran Blair's bequest.

We have established a Board of Trustees for the Museum, six members from The 99s, and three museum experts who live in Oklahoma City. The 99s are, at this time, Anita Lewis, Gene Nora Jessen, Susie Sewell, C.J. Strawn, Jaye Howes, and I. Gene Nora Jessen and Susie Sewell are past presidents of The 99s.

Artifacts, photos, journals, scrapbooks, and other materials have been coming into The 99s Headquarters since the first box arrived in 1969. Verna West and others have carefully cataloged and preserved them. We will be able to exhibit artifacts of women in aviation from the earliest time in the twentieth century to our astronauts.

Now, as this is being printed, the Dedication and Official Museum Open House is scheduled during the International Convention in Oklahoma City in July, 1999 (this year). We hope that all of you will visit our museum sometime. It is right on Will Rogers Airport.

For those of you who want to visit The 99s Website, here is the address: www.ninety-nines.org. The physical address of the museum is 4300 Amelia Earhart Road, Oklahoma City, Oklahoma.

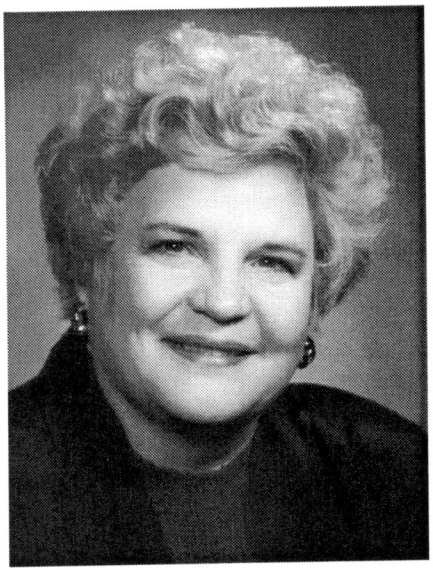

Anita Lewis is chairman of the museum board of trustees.

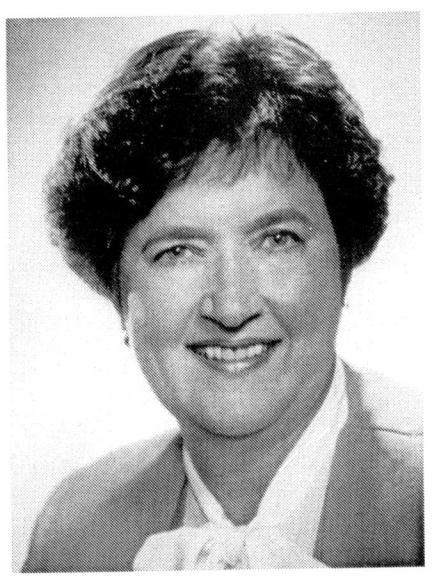

Gene Nora Jessen, 99s President, 1988-1990, is a member of the museum board of trustees.

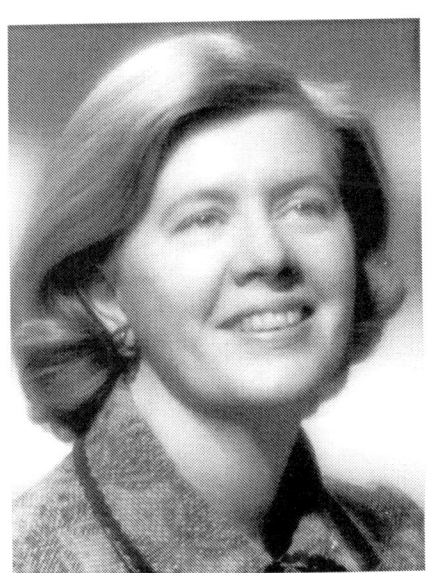

Elizabeth "Susie" Sewell, 99s President, 1972-1974, is on the museum board of trustees.

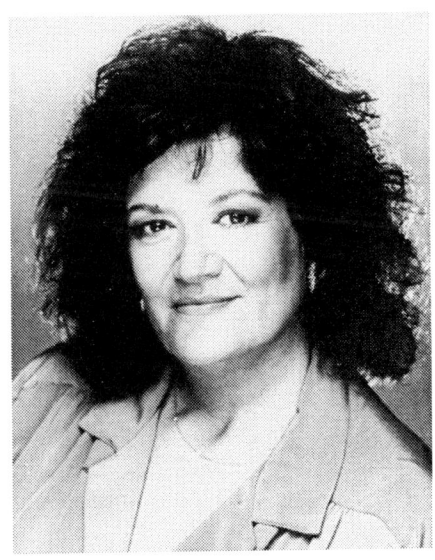

Jaye Howes is a member of the museum board of trustees.

Chapter 20

It is nice to receive honors and awards occasionally on our journey through life. I have received my share, and I have given some to others who have helped me or one of my causes. The finest one given to me was my plaque in the wonderful Forest of Friendship, Atchison, Kansas, in the summer of 1976.

In 1973, our nation started to prepare for the celebration of the American Bicentennial in 1976. The Ninety-Nines wanted to share in the celebration. Our whirlwind known as Fay Gillis Wells took on the job of developing a workable plan. She divided the project into three parts: history, USA in 1976, and new horizons.

Born in 1908, a power and glider pilot since 1929, she wanted to give her project the aviation flavor, one that could be maintained well, and grow each year for a long time.

She is a Charter Member of The Ninety-Nines, and had actively worked alongside other woman pilots to form it into a viable organization. Not an easy task for those involved; each was an assertive person with definite ideas.

Atchison, Kansas, has a special meaning to most women pilots, at least to those of us who were born and learned to fly in the first half of the twentieth century. Amelia Earhart was born in the home of her maternal grandparents, July 24, 1897, in Atchison. The two-story home is now a museum, owned by The 99s.

Fay Gillis Wells envisioned the gently rolling land around Atchison as a park, where granite plaques bearing the names of aviation pioneers could be inset into walks meandering among trees representing every state as well as the 39 countries where there are members of The 99s. The trees would continue to grow and provide shade for the many visitors coming there each year. She called it The Forest of Friendship.

The vision became a reality. In 1976 the Forest was a gift to America on her 200th birthday from The Ninety-Nines, the City of Atchison, and Kansas State University's Forestry Extension.

Forty-two persons were in the first group to be honored, and to my delight, I was among them. I found my plaque near the circle honoring Amelia Earhart, and that was wonderful.

My sister Betty was in the group in 1978 to be honored. Today, there are around 1000 plaques inset in Memory Lane. These names represent aviation from every aspect from its past to its future by honoring those who have, or who still are, contributing to all facets of the advancement of aviation.

This is a magnificent accomplishment, and all due to the vision and energy of Fay Gillis Wells. This is just the latest in a lifetime of accomplishments. She has been the subject of many articles over the years of her long and fruitful life, and her biography may be published soon after you read this. It should be a great adventure story about a truly amazing woman.

Joyce Wells, 99s president from 1994-1996, gave to Gene Nora Jessen and me, the President's Awards for Outstanding Service, for a lifetime contribution to The 99s. Gene, a past president of The 99s, is a member of our museum board of trustees.

In 1997, C.J. Strawn and I received the "Award of Inspiration" from The 99s Board for our "dedication and efforts to see the completion of the museum at Headquarters become a reality."

In February, 1998, Monie Pease began the effort involved in nominating me for one of the National Aeronautic Association's Elder Statesman Awards for 1998. She needed my sister Betty's help in providing certain information and photos, and swore her to secrecy.

I had absolutely no clue that anything was going on until the day in October when the new president of the NAA, Donald Koranda, called me at noon to tell me I had been selected for the award.

In Los Angeles, noon is the favorite time for telemarketers to call. I thought, okay, being selected for an award usually means a cruise on the Caribbean fighting fire, hurricanes, and pirates. And all I have to do is buy a $2000 round trip on the airlines to Florida. Then Mr. Koranda said "Elder Statesman" and "Washington, D.C." I asked some questions and learned my long-time friend, Monie Pease, had nominated me for an Elder Statesman Award for 1998.

Nominations for this annual award had been requested from hundreds of aviation organizations all over the United States. There were 49 nominations given to 25 judges. Seven were chosen, six men and me.

I called Monie and asked her about it. She called Koranda and then called me back and 'fessed up to what she had done.

I could not spare the time and expense to go to Washington, D.C., to attend the ceremony with the six men who had also won the Elder Statesman Award. Monie knew that the Aero Club in Los Angeles was holding its meeting in November and other awards were being presented at that time, so we agreed on that time.

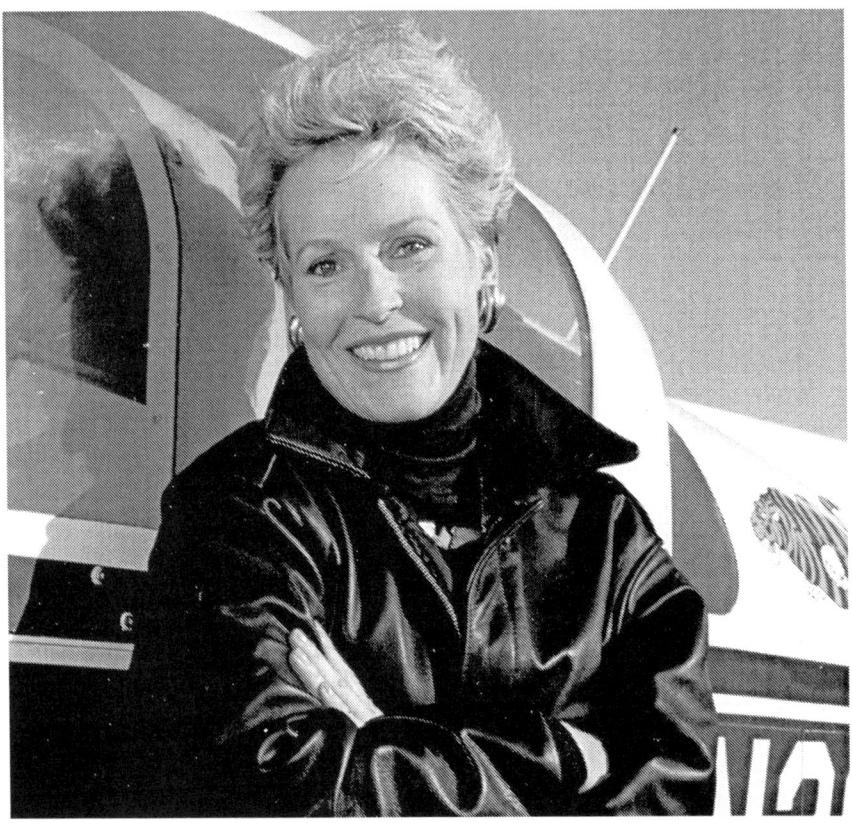

Monie Pease successfully nominated Claire for the prestigious NAA Elder Statesman Award for 1998. In 1991, Monie had done the same for Evelyn "Bobbi" Trout. Monie is a super pilot; she holds several records.

Photo by Richard McGregor. © Richard McGregor

Each year, the NAA presents to five to seven persons the Elder Statesman of Aviation awards. It means elder, too: individuals nominated must be at least 60 years old. It is given in recognition of "significant and enduring contributions over the years to the progress of aeronautics, and demonstrated qualities of patriotism, integrity and moral courage worthy of emulation." The Award was established in 1954.

I take pride in the fact that I met or knew all but three of the 15 women listed since 1955, and many of the 191 men, including the four-star general who stole my drink from the dinner table 40 years ago.

Monie Pease had nominated another winner in 1991, Evelyn "Bobbi" Trout, and I suspect she will continue to watch for others. She's that kind of person.

Monie is quite a flyer. Some years ago she piloted a Grumman Tiger on a 3000-mile flight from San Diego, California, to Caribou, Maine, and set four NAA-approved speed records over the recognized courses: San Diego to Las Vegas, NM; Las Vegas, NM, to St. Louis, MO; St. Louis, MO to Albany, NY; Albany, NY, to Caribou, ME. Three of the four legs included a refueling stop which added to the elapsed time of the record.

Monie was having lunch one day with Bobbi Trout when she, Monie, mentioned that she wanted to fly the diagonal mentioned above. Bobbi suggested that Monie contact the NAA about setting records during that flight. Monie had been thinking about maybe setting some records someday.... And that's how some ideas suddenly jell, with the above result.

In receiving the Elder Statesman Award, I joined a distinguished group. I recognized many of those who had received it through the years: Glenn L. Martin, Igor I. Sikorsky, Richard E. Byrd, William T. Piper, Donald W. Douglas, Sr., Betty H. Gillies, Fay Gillis Wells, James Doolittle, Evelyn "Bobbi" Trout, Jean Ross Howard, Doris E. Lockness, and so many others that it would take up all the space left in this final chapter.

Receiving the award in Los Angeles near my home was an added bonus. Forty to 50 friends and family were able to be present.

Wingspan Air and Space Channel of Bethesda, Maryland, prepared a video documentary of the seven winners for the NAA. They were kind enough to send me a copy.

I do not know what the future holds for me, anymore than anyone else does for themselves. My plans include continuing to gather more funds for the museum's maintenance and operating trust. Building a museum is one thing, but keeping it open and viable is another. We want it to be a place where everyone of all ages will come and be inspired. Who knows when a thought will strike someone that will send the world spinning into another great adventure?

I want to thank all of you I have met and become friends with through the years, for your help in our activities, for sharing the joys and the sorrows, and all that we have accomplished.

We hope you have enjoyed this story of our 75 years of true adventure in this first century of powered flight!

Bibliography

NOTE: Many of these books may be out of print. Check your libraries and used book stores.

Bartels, Diane. *Sharpie: The Life Story of Evelyn Sharp — Nebraska's Aviatrix.* Dageforde Publishing, Lincoln, NE. 1997.

Bell, Elizabeth. *Sisters of the Wind, Voices of Early Women Aviators.* 1920s-30s. ISBN 0-9623879-4-0. 1994.

Bird, Nancy. *My God! It's a Woman.* Australia: P.O. Box 136, St. Ives 2075, N.S.W. Australia. Angus & Robertson, imprint of Harper/Collins. 1990.

Brick, Kay, editor. *Powder Puff Derby, The Record, 1947 to 1977.* Oklahoma City, OK: All-Woman Transcontinental Air Race, Inc. 1985.

Brisseur, Dee (Major, CAF Retired; World's first female CF18 Jet Fighter Pilot). *Achieve it! A Personal Success Journal.* Canada: Unlimited Horizons.

Buroker, Gladys Dawson, *Wind in My Face.* P.O. Box 1093, Rathdrum, ID 83858. 1997.

Churchill, Jan. *On Wings of War, Biography of Teresa James, WAFS/WASP.* Kansas: Sunflower University Press, Manhattan, KS.

Cochran, Jacqueline, and Maryann Bucknum Brinley. *Jackie Cochran.* New York; Bantam Books, Inc. 1987.

Cooper, Ann L., with Dorothy Swain Lewis. *How High She Flies, Dorothy Swain Lewis.* Aviatrix Publishing Company. P.O. Box 485, Arlington Heights, IL 60006-0485. 1999.

Cooper, Ann L. *On The Wing, Jessie Woods and the Flying Aces Air Circus.* Black Hawk Publishing Company. P.O Box 24, Randolph, NJ. 1993.

Douglas, Deborah G. *United States Women in Aviation 1940-1985.* Washington D.C.: Smithsonian Institution Press. 1991.

Harris, Grace. *West to the Sunrise.* Ames, Iowa: The Iowa State University Press. 1980.

Jaros, Dean. *Heroes without Legacy. American Airwomen, 1912-1944.* University Press of Colorado.

Keil, Sally Van Wagenen. *Those Wonderful Women in Their Flying Machines.* New York: Four Directions Press. 1990.

Lomax, Judy. *Women of the Air.* New York: Dodd Mead. 1987.

Marshall, Sheri Coin. *One Can Do It. A How-To Guide for the Physically Handicapped.* Autobiography of a one-armed pilot. (Printed in large type.) ISBN 1-56825-002-9. Rainbow Books.

Moolman, Valerie, and the editors of Time-Life Books. *Women Aloft.* Alexandria, VA: Time-Life Books. 1981.

Morrissey, Muriel Earhart, and Carol L. Osborne. *Amelia, My Courageous Sister.* Santa Clara, CA: Aviation Archives. 1987.

Ninety-Nines, Inc. *The Ninety-Nines — Yesterday — Today — Tomorrow.* Turner Publishing Company. Paducah, KY. 1996.

Oliver, Susan. *Odyssey, A Daring Transatlantic Journey.* New York: Macmillan Publishing Company. 1983.

Otypka, Sylvia J. *On Becoming an Airline Pilot.* Leading Edge Publishing Company, P.O. Box 461605, Aurora, CI 80046. 1997.

Powell, Mary Curtner. (J-9 to 14.) *Queen of the Air, Story of Katherine Stinson.* Dallas: Coldwater Press, Dallas, TX 75228.

Prince, George. *Mama Bird, A Biography of Evelyn Bryan Johnson, A Flight Instructor.* Evelyn Bryan Johnson, P.O. Box 1013, Morristown, TN 37816.

Render, Shirley. *No Place for a Lady.* Canada's female aviators from 1928 to 1994. Canada.

Rich, Doris. *Bessie Coleman: Daredevil Aviator.* Publisher: SI Press.

Sacchi, Louise. *The Happy Commuter.* Sacchi delivered over 300 aircraft around the world. Eastern Pennsylvania Chapter 99s, c/o Mary Wunder, 743 Collegeville Rd., Collegeville, PA 19426.

VanScyoc, Mary Chance. *A Lifetime of Chances.* One of United States' first female air traffic controllers. Parkwood Press, P.O. Box 20550, Wichita, KS 67208-1550. 1997.

Veca, Donna, and Skip Mazzio. *Just Plane Crazy: Biography of Bobbie Trout.* Santa Clara, CA: Osborne Publisher. 1987.

Wagstaff, Patty, with Ann L. Cooper. *Fire and Air, A Life on The Edge, Patty Wagstaff, Aerobatic Champion.* Chicago Review Press, 814 North Franklin Street, Chicago, IL 60610. 1997.

Whyte, Edna Gardner, with Ann L. Cooper. *Rising Above It, An Autobiography, Edna Gardner Whyte.* New York: Orion Books. 1991.

MAJOR RESOURCE OF PUBLISHED MATERIAL ON WOMEN IN AVIATION:

Niekamp, Dorothy. *Women and Flight, 1910-1978: An Annotated Bibliography.* And its update: *Women and Flight, 1978-1989.* Oklahoma City, OK. Publisher, The Ninety-Nines, International Organization of Women Pilots, Inc. 1990.

MAGAZINE RESOURCES:

International Women Pilots/99 News, non-member rates available. The Ninety-Nines Inc., Box 965, 7100 Terminal Drive, Oklahoma City, OK 73159.

Woman Pilot Magazine, P.O. Box 485, Arlington Heights, IL, 60006.

Index

A-B
Ahrens, Peter: 133-167
Albani, Helen: 83
Aldrich, Clare: 81
Amelia Earhart Memorial Race: 34, 52
Armstrong, Ann: 192
Baker, California: 16
Banks, Marian: 54
Barstow, California: 122
Barton, Rikki: 182, 184, 201
Bastian, Rhea and Richard: 186
Bateman, Justine: 193
Beard, Melba Gorby: 41
Beck, Ethel: 131
Becker, Mahlone: 188
Becker, Robin: 188, 192, 209
Beinhorn, Elly: 77-90
Bera, Frances "Fran": 3, 53-56, 59, 76, 83, 87, 110
Berry, Margaret Ross: 171, 179
Bird-Walton, Nancy: 77, 81-82
Bishop Airport, California: 43
Bixby, Dianna Converse Cyrus: 33, 35-36, 62
Blair, Mary Francis "Fran": 192, 205, 209-210
Boot, Barbara: 187
Bower, Edna: 83, 87
Brach, Fred: 15-16
Brick, Kay: 34
Brown, Fidge: 188
Brown, Lori: 188
Brown, Mitzi: 188
Bruno, Jerry: 71
Buffalo Barbecue Fly-In: 122

C
Callaway, Margaret: 85, 106
Callaway, Lt. Colonel Richard: 85
Camden, Conrad: 181, 191
Camden, Lillian: 182, 191
Campbell, Mary Ann: 185
Campbell, Pete: 96
Carl, Jann: 193
Cassell, Jerelyn: 90
Choi, Pam: 193
Claire Walters Flight Academy: 91, 106
Cochran, Jacqueline: 14-15
Colapietro, Leo J.: 72
Colvin, Joe and Pat: 130-132
Conant, Roger: 72
Conner Field, Quartzsite, Arizona: 7-8
Conrad, Max: 80, 147
Cooper, Ann L.: 57
Cranford, Bill: 29
Crawford, Carolyn and Sharon: 185
Crews, Nancy: 103
Critchell, Iris: 76-77, 83
Crunk, William J.: 71
Cummings, Dodie Prario: 54

D-E
D'Armand, Ann Savage: 181, 193
Danison, Leonard W.: 72
Davis, Charles M.: 95
Davis, Clara: 53
Dick, Helen: 33, 50
Dietrich, Jan and Marion: 57
Dinan, Elizabeth: 15, 109
Earhart, Amelia: 2-3, 15, 156, 212
Edwards, Steve: 10
Elder Statesman Awards: 213
Ettridge, Douglas: 208
Ewert, Margaret "Peggy": 208

F-G
Failing, Joyce Carl: 22, 122-123, 180
Farrar, Les: 20, 25
Farrar, Robert: 20
Faux, Betty: 95, 115, 131, 133, 163
Fields, Summer: 112
Fischbeck, Dr. George: 193
Forest of Friendship: 212-213
Fox, Michelle Brittingham: 99, 189
Fullerton Air Service: 27
Funk, Shelly: 112
Futterman, Norma: 189, 200
Fyn, Adriana: 81
Gammell, Hank: 130
Gammell, Jan: 196
Gavin, John Anthony Golenor: 68, 69, 121
Getty, Estelle: 193
Gibbs, Ellsworth: 15
Gillies, Betty Huyler: 15, 57, 66
Glasson, Pauline: 93
Glenn, William: 169
Graham, Dr. Ralph: 61
Grandy, Mary Jo: 185
Griffith, Thon: 179
Grill, Ralph: 9
Grupenhagen, Esther: 190

H
Hackett, Gene: 122
Hagedon, Harriet: 130-132
Hamilton, Dolores Pynes: 188
Hartwick, Don D.: 71
Hays, Robert: 122, 193
Herd, Helen: 85
Herrin, Billy: 180
Hewitt, Helen: 190
Hicks, Betty: 91
Howes, Jaye: 206, 208, 210
Hughes, Prince: 193

I-J-K-L
Ives, Vera: 188
Jessen, Gene Nora: 210, 213
Jugend, Harry: 192, 205
Kass, Gail: 201, 209-210
Koich, John: 179-180
Koranda, Donald: 213
Krantz, Judith: 182, 193
Laing, Diann: 194-196
LaMar, Jane: 113, 187

Laughbaum, Marilyn: 192
Laughbaum, Nina: 182, 192
Laymon, Lois: 86
Leatherwood, Shirley: 187
Leerskov, Dixie C.: 12
LeFevre, C.A. "Al": 13
lenticular clouds: 44-47, 51
Lewis, Anita and Vic: 206, 210
Llorens, Maribel: 190
Loufek, John E.: 32, 44, 49, 51

M

Macafee, Robert: 114
MacCready, Paul B., Jr.: 44, 48
Macofsky, Stuart: 193
MacPherson, Beatrice Edgerly: 178
Madden, Cheryl: 188
Majors, Ethel: 15
Mauch, Gene: 126
McCreery, Ardath: 83
McGuire, Emma: 91, 99, 105, 192
McKendry, Felicity: 81
McLaughlin, Emily: 193
McMillen, Deane: 21, 32
Mead, Margaret: 180
Medes, Beatrice "Bea": 34, 52
Merrifield, Rosalind: 133-167
Mexico Mercy Flight: 130-132
Mickelsen, Gerry: 87
Miller, Betty: 133
Minter, Doris: 105, 192
Moisant, Mathilde: 201
Moskow, Gene: 42
Most, James L.: 41, 54, 56
mountain wave: 32, 44, 47, 51

N-O-P

nation's first airline captain: 35
Newell, Juanita: 90
Neuman, Golda: 208
Niekamp, Dorothy: 210
Ninety-Nines, Inc. The: 14-15, 201-213
Noonan, Frederick: 156
Oliver, Susan: 180, 193
Owens Valley, California: 19-20, 43, 47
Palmer, Katherine: 81
Palm Springs Airport, California: 39
Palms to Pines Air Races: 110, 112, 179-199
Pease, Monie: 213-214
Pickering, Aileen: 189, 200
Potter, Ilovene: 196
Powder Puff Derbys: 3, 34, 52, 77, 91
Prigmore, Susan: 187
Prineville, Oregon: 185

Q-R

Quartzsite, Arizona: 8, 43
Reitsch, Hanna: 45, 51
Rey, Michael: 167
Ringenberg, Margaret: 86
Romero, Beverly: 188
Rose, Jean Parker: 54, 76
Ross, Harland: 43-44, 51
Ryan, Bertha M.: 50-51
Rungling, Dorothy: 81

S

Sanders, Darline "Dottie": 54
Santa Ana High School: 3, 20, 148
Santa Monica Airport, California: 31, 52, 63, 71, 104, 112
Saunders, Aileen: 90
Schiffmann, Charles "Pat": 188
Schiffmann, Jean Larson: 22, 180, 188, 196-199
Schilling, Lisa: 194
Schumaker, Max: 75
Scott, Eleanore: 188
Scott, Sheila: 147
Sewell, Elizabeth "Susie": 210
Seymour, Boots: 54
Sharp, Art: 54
Sheehy, Ethel: 14-15
Shelton, Jimmye Lou: 90
Sherman, Karen: 180, 196
Shock, Margaret "Maggie': 59
Silver Lake Airport, California: 16, 22
Smith, Joan Merriam: 133
Strawn, C.J.: 201-210, 213
Strudwick, Peter: 68
Symons, Robert: 43-44, 51

T

Taylor, Buck: 185
Taylor, Donna: 113, 187
thermal soaring: 32, 40, 44
Theurkauf, Susan: 202, 207
Thisted, Barbara: 78, 83
Thomas, Ralph: 134
transpacific solo flight: 133-167
Trout, Evelyn "Bobbi": 214
Tucker, Barbara and John: 173

V-W

Van Zelf, Lois: 187
Vasko, Donna: 184
Wagner, Ruth: 87
Walker, Dororhy: 188
Waller, Marta: 193
Walters, Michael (Sr.): 55, 57, 59, 60, 63
Walters, Michael (Jr.): 63, 127, 134
Walters, Susan: 63, 99, 134
WASPs—Women's Airforce Service Pilots: 6, 8, 14
Weldon, Cecilia and Jeff: 192
Wells, Fay Gillis: 212-213
Wells, Joyce: 203-204, 207, 213
West, Carolyn: 33-34, 52
West, Verna: 210
Weyle, John A.: 201
White, Frosty: 179
Whyte, Edna Gardner: 57
Williams, David and Diane: 183
Willmore, Winifred: 81
Wolters, Claire: 147
Woodward, Betsy: 33, 50

Z

Zeitlin, David: 201
Zerbe, Anthony and Jared: 122

Please make a copy of this page to order additional copies.

Ordering Information

To order additional copies of
This Flying Life by Claire L. Walters
and Betty McMillen Loufek,
please fill out the form below and
send check or money order to:

AirWoman Press
P.O. Box 721
Camarillo, CA 93011-0721

Per book, $19.95 plus $3.00 postage
California residents, add $1.45 for tax.

SEND TO:

_____
Name

_____
Address

_____
City/State/Zip

_____
Phone number

_____
Number of books

_____
Total amount enclosed

Thank you for your order!